Taunton's

BUILD LIKE A PRO™
EXPERT ADVICE FROM START TO FINISH

Building
a Shed

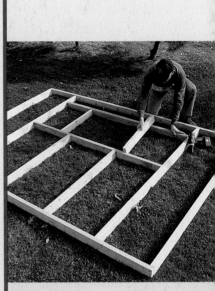

Taunton's

BUILD LIKE A PRO™
EXPERT ADVICE FROM START TO FINISH

Building a Shed

Joseph Truini

The Taunton Press

The Taunton Press
Inspiration for hands-on living®

The Taunton Press, Inc., 63 South Main Street, P.O. Box 5506, Newtown, CT 06470-5506
e-mail: tp@taunton.com

EDITOR: Jennifer Renjilian Morris
COVER AND INTERIOR DESIGN: Kimberly Adis
LAYOUT: Cathy Cassidy
ILLUSTRATOR: Mario Ferro
PHOTOGRAPHER: All photographs by Geoffrey Gross, except the following: p. 6: © Joseph Truini; built by Better Barns; p. 7: (top) © Joseph Truini; built by Better Barns; (bottom) © Joseph Truini; p. 8: © Joseph Truini; built by Better Barns; p. 9: © Joseph Truini; p. 10: (left) © Joseph Truini; (right) © Smith-Baer; pp. 13–14: © Joseph Truini; pp. 15–16: Walpole Woodworkers; p. 17: © Jerry Kolesar; p. 18: © Crandall & Crandall; p. 19: © Saxon Holt; p. 21: Walpole Woodworkers; pp. 22–23: © Joseph Truini; p. 28: (bottom) © Joseph Truini; p. 29: Photo courtesy DekBrands®; p. 32: Photo courtesy SFPA; p. 38–39: © Smith-Baer; p. 42: © Smith-Baer; p. 46: (bottom) Photo courtesy Lee Valley Tools; pp. 48–50: © Joseph Truini; p. 52: (middle and bottom) © Joseph Truini; p. 53: (left and top right) © Joseph Truini; (bottom right) Walpole Woodworkers; p. 56: (top) © Joseph Truini; (bottom) © Western Red Cedar Lumber Association; p. 57: © Western Red Cedar Lumber Association; p. 58: (top left) © Joseph Truini; (top right) © Western Red Cedar Lumber Association; p. 59: © Western Red Cedar Lumber Association; p. 60: © Cedar Shake and Shingle Bureau; p. 61: (bottom) © Georgia Pacific; p. 62: © Georgia Pacific; p. 63: (top and middle) © James Hardie® Building Products; (bottom) © Weatherboard; p. 64: © Joseph Truini; p. 65: (middle) © Cedar Shake and Shingle Bureau; (bottom) © Joseph Truini; pp. 66–70: © Joseph Truini; pp. 72–75: © Joseph Truini; p. 78: © Joseph Truini; p. 104: Photo courtesy Handy Home Products; p. 105(top and bottom left) Photos courtesy Outback Wood Products; (right and middle left) Photos courtesy LancasterBarns.com; p. 122: © Joseph Truini, built by Better Barns; p. 123: (top left) Photo courtesy Georgia Pacific; (middle left) Photo courtesy Handy Home Products; (bottom left) © Joseph Truini; (right) Photo courtesy Walpole Woodworkers; p. 152: (top) © Joseph Truini; (bottom) Photo courtesy Walpole Woodworkers; p. 153: (top left, bottom left, and top right) © Joseph Truini; (middle) Photo courtesy Walpole Woodworkers; p. 184: (top) Photo courtesy Walpole Woodworkers; (bottom) © Joseph Truini; p. 185: (top) Photo courtesy Walpole Woodworkers; (middle and bottom) © Joseph Truini

Library of Congress Cataloging-in-Publication Data
Truini, Joseph.
  Building a shed. -- Completely rev. and updated.
     p. cm.
  Includes index.
  ISBN 978-1-56158-966-1
  1. Sheds--Design and construction--Amateurs' manuals. I. Title.
  TH4962.T78 2008
  690'.89--dc22

                          2008020828
Printed in the United States of America
10 9 8 7 6 5 4 3 2 1

Construction is inherently dangerous. Using hand or power tools improperly or ignoring safety practices can lead to permanent injury or even death. Don't try to perform operations you learn about here (or elsewhere) unless you're certain they are safe for you. If something about an operation doesn't feel right, don't do it. Look for another way. We want you to enjoy the craft, so please keep safety foremost in your mind whenever you're in the shop.

To my father, Joseph Truini Sr., who taught me that working with your hands—whether building a house or writing a book—is a noble pursuit.

To Marla, Kate, and Chris, for keeping me afloat.

ACKNOWLEDGMENTS

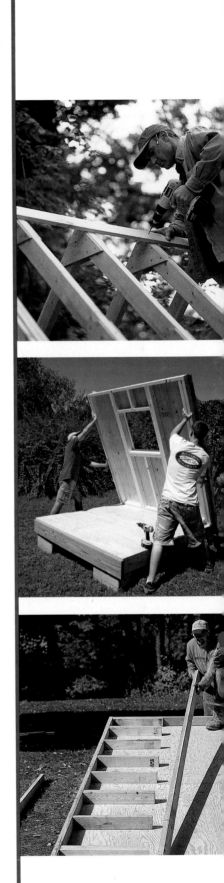

Writing a book is a lot like building a shed: It always takes longer than you think, but in the end, it's well worth the effort. It's also virtually impossible to do without the help of others.

Thank you to all the hard-working, talented people at The Taunton Press. I'm proud to have my name on one of your books. Thanks especially to executive editor Helen Albert, who first offered me this book project and showed confidence in my ability from the very beginning. I'm also grateful to Wendi Mijal for her skilled art direction, and Nicole Palmer for her always friendly, helpful assistance.

Thanks, too, to book editor Jennifer Renjilian Morris, for her keen eye, helpful suggestions, and professionalism throughout the arduous task of molding my words.

I was extremely fortunate to have worked with photographer Geoffrey Gross, whose technical skill, immeasurable patience,

and dedication to this project can't be overstated. The words may be mine, but it's Geoffrey's images that truly make this book special.

I could never have produced this book without the help and unflagging support from Pete and Arlene Charest, owners of Better Barns, a Connecticut-based shed-building company. They were always open to suggestions and showed an inordinate amount of patience throughout every design and construction phase of this project. Thanks, too, to their talented carpentry crew, Greg Butkus, Paul Clough, and Dave Scarles, for never running out of patience or good humor.

Finally, a debt of gratitude is owed to John Blaney and Bill Jenks, building officials in Litchfield County, Connecticut, who expertly and patiently answered all my arcane (and sometimes inane) questions about building codes, zoning regulations, and frost heave. Their contributions made this a much better book.

CONTENTS

INTRODUCTION

As the son of a carpenter, I had no choice but to learn my father's craft. It was passed down to me through the same complex genetic code that gave me his dark eyes and stubborn nature. My apprenticeship began at the tender, clueless age of 12, when my father woke me early one summer morning (much to my surprise) and announced that it was time to go to work. As I climbed into his truck, sleepy-eyed and silent, I had no idea that I was embarking on a journey of discovery and purpose that would guide me through life to this very day.

During that summer, and subsequent summers for the next several years, I worked alongside my dad and learned a lot about carpentry, electrical wiring, plumbing, and life. Later, as a young adult, I used those skills to earn a living as a remodeling contractor, union carpenter, cabinetmaker, and, eventually, writer.

For the past 26 years, I've written about homebuilding, remodeling, and woodworking for nearly every possible medium: books, newspapers, websites, television, radio, and several magazines, including *This Old House, Popular Mechanics, Today's Homeowner, Country Living, Woman's Day, Handy,* and *Home Mechanix.* It was during my years as a magazine editor that I first began writing about backyard outbuildings and discovered that nearly every homeowner—regardless of the size of his or her house—needed a storage shed. Over the years, I've written articles about and built more than a dozen sheds. Therefore, when the fine folks at The Taunton Press asked me to write this book, I was confident that my experiences in the building and publishing fields would prove invaluable. I hope you think so, too.

From the very beginning, the goal of this book has been to provide you with the information, inspiration, and confidence to build your own backyard storage shed. And not just one of the structures shown in this book, but any shed at all. Like most other shed books, this one has a lot of photos of attractive outbuildings and many construction drawings, but that's where the similarities end. This book gives specific information on what you need to know before you build a shed, including design considerations, building code issues, and evaluating your storage needs. A chapter devoted to shed-building techniques covers everything from foundations to roof framing. Another chapter explains the wide variety of shed-building materials, including siding, roofing, doors, and windows.

However, what makes this book truly unique is that each of the final five chapters shows how to build a particular shed from scratch. These structures were specifically designed and constructed for this book and include an easy-to-build 2-ft. by 6-ft. Lean-to Shed Locker that's perfect for storing lawn and garden tools; a modest 6-ft. by 6-ft. Garden Shed that has a stylish hip roof and fiber-cement siding; a charming 8-ft. by 12-ft. Saltbox Potting Shed, which features a cedar-shingle roof and a traditional Dutch door; a 10-ft. by 16-ft. Colonial-Style Shed that combines classical design with beautiful vertical-board cedar siding; and a spacious 12-ft. by 20-ft. Gambrel Storage Barn, which has double-wide sliding doors and an interior storage loft.

For each shed, there are dozens of step-by-step photos and precise drawings to guide you through each phase of the construction process. However, if you need additional information, mail-order plans are available for each of the five sheds. In this edition we've also included Design Options at the end of each chapter. There you'll find different looks that go beyond the construction basics of the text and show you what else is possible.

It took about a year to produce the book you now hold in your hands, but it—and all that I am today—really started early one summer morning more than 40 years ago.

BEFORE YOU BUILD

Sooner or later, most of us settle into a home that comfortably accommodates our lifestyle, one with enough bedrooms and bathrooms and perhaps even an adequate number of closets. But regardless of whether it's a cottage or a castle, we never have enough accessible storage space for all of our stuff. And as we acquire new things, our living spaces become increasingly cramped and crowded.

Basements, attics, and closets are fine for keeping holiday decorations, luggage, boxes of books, and old clothing, but a separate outbuilding is the only practical solution for storing and organizing garden tools, lawn mowers, camping gear, bicycles, ladders, lumber, and all the other things we typically cram into the garage or pile up outdoors.

The challenge, however, isn't simply to build a shed, it's to build the right shed. With a little forethought and planning, it's easy to determine which size and style shed is best for you and your family. And you can begin by answering this simple question: What are you going to use the shed for? ▶ ▶ ▶

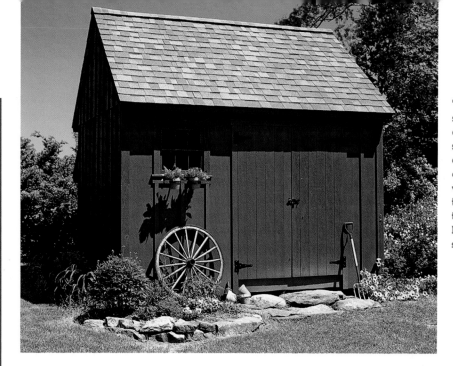

The generously sized doors on this all-purpose storage shed provide easy access to the shed's contents. Vertical wood strips are nailed to plywood siding to give the look of board-and-batten siding.

PRO TIP

Make a list of all the items and pieces of equipment that you want to store in your shed. Then measure the largest items and size the building accordingly.

TRADE SECRET

To better visualize how a shed will fit on your property, measure the four corners of the shed, drive a stake into the ground at each corner, then stretch a string between the stakes to represent the shed's "footprint." Imagine how the shed will look, keeping in mind that it will seem bigger in winter, when the leaves are off the trees.

Evaluating Your Storage Needs

The first step in building a shed is to figure out how the shed will be used and which items will be stored in it. I know it's tempting to skip this step and jump right to the fun part: pounding nails. But taking the time now to assess your storage needs will help you avoid a situation I like to call transfer dumping. That's when people impulsively build or buy a shed—any shed—just so they can transfer stuff from the garage and dump it in the shed. The result, not surprisingly, is a cluttered, impenetrable mess. Remember, a shed should offer well-organized, accessible storage. You should be able to walk in and retrieve an item without fear of being buried alive.

Selecting the right size shed

There's no such thing as a "standard" size storage shed. I've seen them as small as 2 ft. by 4 ft. and as large as 16 ft. by 20 ft. (to me, anything bigger is a barn). Determining which size shed is best for your needs isn't always an easy task, but you can gain some insight by answering the following questions:

ACCORDING TO CODE

Don't be too surprised if you discover that local zoning ordinances or deed covenants prevent you from building a shed. An increasing number of planned communities have banned outbuildings altogether because they consider them to be eyesores. This "beautification" act is typically enforced in thickly settled neighborhoods with postage-stamp-size lots.

• How much available yard space do you have?
• Are there any size restrictions imposed by the local building department, zoning board, or homeowners' association?
• How far from the property line must the shed be built?
• Will you need to store very long items, such as extension ladders or lumber?
• How big is your budget?

Appearance is also important when sizing a shed. Be aware that its proportions should complement the available space and surrounding landscape. Avoid the all-too-common mistake of building a giant shed in a tiny backyard. It'll visually overpower the entire property, including your home.

It's also smart to build a shed to even-number dimensions (i.e., 8 ft. by 10 ft., 10 ft. by 12 ft., 12 ft. by 16 ft.). That will greatly reduce waste by taking full advantage of common building materials, such as plywood and framing lumber, which come in even-number sizes and lengths.

Before you decide on a particular size shed, contact your local building department. It'll be able to alert you to any restrictions regarding the shed's minimum and maximum square footage (width times length). Check, too, on the maximum allowable height. Ask the building inspector for a list of documents that you'll need to apply for a building permit. (For more information on building code requirements, see p. 18.)

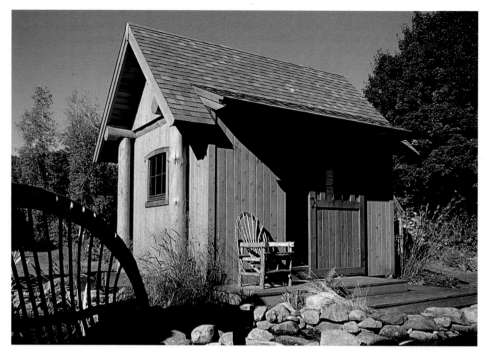

This elegant Adirondack-style cabin features an exposed frame made of 10-in.-dia. red-pine logs. Measuring 12 ft. by 12 ft., it's large enough to be used as an artist studio, a garden shed, or a pool cabana.

Design Considerations

Storage sheds are, by their very nature, utilitarian. They're often looked upon as nothing more than repositories for our stuff. But that doesn't mean they can't also be well-designed, solidly built, and architecturally interesting. Unfortunately, most of the prefabricated sheds sold at home centers and lumberyards don't fall into any of those categories. Take a drive through any neighborhood and you'll likely find the landscape dotted with rusty, leaning metal buildings and wobbly wooden shacks that appear to be on the verge of collapse. (I recently saw a tin-can shed so badly rusted that the only thing keeping it up was the stack of lumber stored inside. Every time a board was removed, the shed slumped a little lower.) That's why building your own shed is the best way to get a storage building that's functional, durable, and an asset—not an eyesore—to your property.

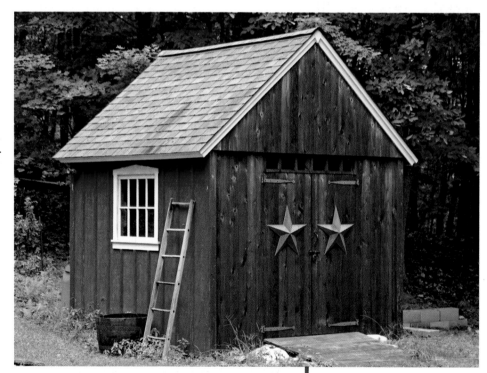

The rustic appearance of this outbuilding is due in part to its painted sides and natural-finish front. This 10-ft. by 12-ft. storage shed was built to even dimensions to take advantage of standard, even-size lumber.

This elegant English potting shed has a small porch, vertical-board siding, and faux-slate roof shingles. Western-red-cedar siding was chosen for its good looks and long life.

PRO TIP

Neither aluminum nor vinyl siding is rugged enough to absorb the inevitable hard knocks and rough handling that every outbuilding endures.

TRADE SECRET

Can't decide on which type of siding to use for your shed? Call a local lumberyard and get price quotes for each type. Costs vary widely, and price alone may be the deciding factor. Remember: You can't build a redwood shed on a plywood budget. Availability also plays a key role, which is why it's important that you place your siding order early. Lumber, like produce, isn't always available every day of the year. Your construction schedule can easily be delayed by two or more weeks if the siding you desire is suddenly unavailable.

Identifying style

Once you have a rough idea of what size shed you'd like to build, you need to decide on its architectural style, which is often dictated by the roof design and siding material. Most sheds have a simple gable roof, though saltbox, gambrel, and shed roofs (like the roof on the Lean-to Shed Locker in Chapter 4) are other options.

When it comes to siding the exterior walls of a shed, you also have several options. Wood is by far the most commonly used material, which isn't surprising, as it's readily available, affordable, and easy to work with. Choices in wood include bevel siding (a.k.a. clapboards), shingles, vertical boards, and plywood. Plywood siding isn't quite as stylish as other types of wood siding, but it's less expensive and goes up very quickly, which is why it's often used on larger sheds and barns.

Fiber-cement siding—a relatively new shed-building material—offers a practical alternative to wood clapboards. It looks like wood but is made of fiber-reinforced cement, so it's extremely durable and completely fireproof.

Window wisdom

Windows also influence a shed's style, but to a slightly lesser degree than the roof or siding does. When designing your shed, remember that the size and placement of the windows should be in proportion to the building.

You don't have to be Frank Lloyd Wright to figure out which size windows will look best; all you need to do is open your eyes. Look around your neighborhood or local lumberyard to find a shed that's about the same size as the one you're planning to build. Stand back far enough to view the entire building and surrounding area. Look for a pleasing balance between the amount of glass and the visible siding. The windows should be properly proportioned, neither too big nor too small for the structure. When you find a well-proportioned shed, measure the size and placement of its windows.

Here's another equally effective method that doesn't require you to even leave the house: Get a ruler and make a scale drawing of the building. Then sketch in the windows, adjusting their dimensions until they look well balanced. For example, if two small, square windows don't look right, then enlarge them. If that doesn't help, replace the two windows with a single large, rectangular one.

Maintaining proper proportions applies to all parts of the building, including the exterior trim around the windows and doors. Generally speaking, smaller windows and doors look best with narrow trim—that is, trim less than 3 in. wide.

Large windows and wide doorways can accommodate much wider trim boards.

Interior features

As you're working your way through the initial design and planning stage, think about customizing the interior of the shed, too. Even if you're going to use the shed for only general storage, chances are good that you'll need at least one or two shelves and a few hanging hooks.

A wooden workbench is a popular shed accessory, but build one only if you're sure you'll be able to keep the floor clear of clutter. It's very easy for a shed to become so crammed full of stuff that you can't even see the workbench.

If you plan to store bulk goods in the shed, such as fertilizer, potting soil, or pet food, be sure to allow space for a few large plastic storage bins.

Selecting a style

Choosing a particular shed style is often based on personal preference and taste, but there are also some practical concerns. For example, sheds with gable roofs offer tall sidewalls that are useful for putting up shelves and hanging long-handled tools. The tall walls can also accommodate larger windows and doors. However, the steeply sloping roof doesn't provide much space above the ceiling joists or a lot of headroom near the walls.

Gambrel-style roofs—sometimes referred to as barn roofs—are quite spacious inside and offer plenty of headroom. But their shorter sidewalls offer very little space for hanging tools or mounting shelves. It's also difficult to install doors on the sides of most gambrel-style sheds because of the limited wall height. Some builders solve these problems by building taller sidewalls. That

When building a large shed, consider a door at each end or a pair of doors in the middle of a sidewall. You'll reach items stored all the way in the back more easily.

TRADE SECRET

Looking to add a workbench to your shed but don't quite have the space for one? Try this space-saving solution: Make a 16-in.- to 24-in.-wide by 6-ft.- to 8-ft.-long benchtop out of ¾-in. plywood. Measure 36 in. up from the floor and attach the benchtop to the wall with fold-down metal brackets (available at any hardware store for about $15 apiece). The collapsible brackets allow you to lower the benchtop flat against the wall when it's not in use.

The unusual double-gable roof design of this backyard building creates two separate, yet connected, storage areas. On the left side is a ramp and doublewide doorway; this area is used to store a lawn tractor. On the right, brick steps and a single door lead to a tool-storage area.

This handsome 10-ft. by 14-ft. shed features a post-and-beam frame made with 6×6 timbers; it was built from plans by Mercurial Editorial.

This unique storage barn combines a gable-roof main building with an attached shed-roof extension. The result is a countrified outbuilding with loads of storage space.

way, they gain usable wall space without sacrificing headroom. In fact, many barns are built this way, but they are large enough to accommodate the extra wall height; most sheds are not. Simply building taller walls on a shed will throw the entire building out of proportion and make it look oddly stretched.

A Working Plan

Once you've decided on the size and style of the shed you want to build, you need a building plan. There are two basic ways to obtain one: buy an existing plan or draw one from scratch.

Plan sources

Mail-order shed plans are available from books, magazines, websites, tool catalogs, and companies that sell homebuilding plans (see Resources on p. 214). Such plans typically cost between $10 and $50, depending on the size and complexity of the building. Buying a ready-made plan will save you the trouble of designing and drawing a shed from scratch, which is a rather time-

consuming process. However, drawing your own plan allows you to create a unique shed that's specifically designed to suit your tastes and storage needs; it also costs less. A third option is to buy a mail-order plan, then alter the design to satisfy your requirements.

Looking for a better deal? Many building-product manufacturers, such as Georgia Pacific® and Simpson Strong-Tie®, offer free plans in the hope that you'll then buy their lumber or hardware.

Drawing a design

If you can't buy a plan of the exact shed you'd like to build, you'll have to draw it yourself. Now don't panic. You don't have to be a skilled draftsperson to create an accurate building plan. All you need is some ¼-in.-scale graph paper and a few basic drawing instruments, such as a mechanical pencil, a ruler, dividers, a protractor, a compass, and an eraser.

Start by drawing an elevation view of each exterior wall of the shed. This is simply a flat, one-dimensional view of the wall, as it would look if you were standing directly in front of it (see the drawing on the facing page). Be sure to include the roof and the location of the doors and windows. Draw the shed using a ½-in. scale. In other words, each ¼-in. square printed on the paper is equal to 6 in. That means 1 ft. of actual

Ridge vent

Asphalt shingles

⅝-in. textured plywood siding

Front Elevation
A front elevation is a one-dimensional view of the front façade drawn to scale. It should include the size and location of the doors and windows and indicate the type of roof shingles and siding that will be installed.

2-ft. × 2-ft. window

5 ft.

16 ft.

10-in-12 roof slope

12 ft.

2-ft. × 3-ft. window

⅝-in. textured plywood siding

Side Elevation
A side elevation drawing shows the end of the building and indicates the slope of the roof. If both ends of the shed are identical, you only need to draw one side elevation.

10 ft.

TRADE SECRET

When you find a photograph of a shed that you're interested in building, you need to know only one critical dimension to figure out the rest of the shed's proportions. Lay a ruler across the known dimension, which can be any part of the shed, such as its length, width, or rafter thickness. Note the measurement on the ruler and use that as your scale. For example, if you know the building is 10 ft. wide and it measures 5 in. in the photo, then 1 in. represents 2 ft. of actual space. Knowing which scale to use allows you to calculate other dimensions easily, including door width, wall height, and window size and spacing.

distance is represented by two squares, or ½ in. By the way, you can buy ½-in.-scale graph paper, but the smaller squares in the ¼-in. grid allow you to draw 6-in. dimensions much more accurately.

The reason it's important to draw the elevations to a precise scale is that it's the easiest way to tell whether the shed will be properly proportioned. If any part of the building is even slightly out of whack, it'll be clearly visible in the drawings. Here are some things to look for:

• Does there appear to be too much space between the windows? Enlarge them, add shutters, or install another window in between.

• Is the shallow roof pitch making the shed look a little squashed? Redraw the roof at a steeper angle.

• Does the shed look too long and narrow? Decrease its overall length or make the building wider.

• Don't be afraid to experiment with different designs. Making adjustments to the shed at this point is easy—just pull out your eraser and try again.

After completing the elevation drawings, sketch a plan view (a.k.a. bird's-eye view) of the proposed foundation. Be sure to include the exact spacing and alignment of the foundation

A TYPICAL FOUNDATION PLAN

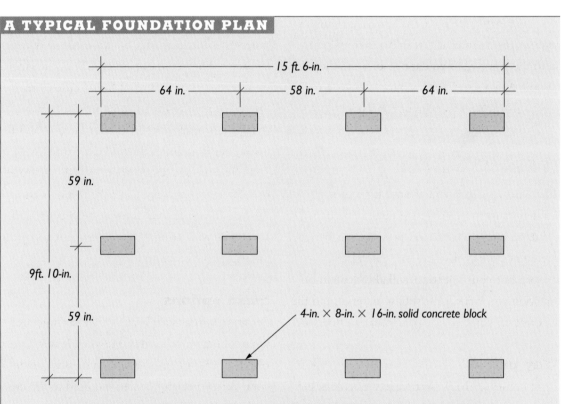

A foundation plan is a bird's-eye view of the proposed foundation. This drawing is of an on-grade foundation made up of 12 solid-concrete blocks. A foundation plan must include the size of the blocks and the spacing between them.

system, which can be concrete blocks, footing holes, wooden skids, or poles. For specific information on foundations and a wide variety of shed-building techniques and materials, see Chapters 2 and 3.

Sweat the details

You typically don't need to draw a separate framing plan of how the floor, walls, and roof will be framed, but include this information somewhere on the drawing. Be sure to list nominal dimensions and names of the parts, along with the layout spacing and sheathing. For example, for the walls you might add the following construction detail: 2×4 wall studs spaced 16 in. on center and sheathed in ½-in. exterior-grade plywood. It's necessary to include all the major construction details because most building departments require you to submit this information along with the plan when you apply for a building permit. Again, see Chapter 2 for various building techniques for each construction phase.

Finishing School

The exterior finish you choose for a shed also influences its style. There are three basic choices: paint, stain, and clear wood preservative. Personal preference plays a key role in which finish you choose, but your selection will also depend on the look you want, your siding material, and the amount of maintenance you want to undertake.

Why paint?

Paint comes in the widest variety of colors and lends a more formal look to an outbuilding. However, keep in mind that painted wood will eventually need to be scraped and repainted, and most people would rather vacuum Texas than scrape paint.

This antique Connecticut carriage house was restored and converted into an attractive storage building. The combination of painted wood and stone veneer add a refined, formal look to the building.

If you're going to paint a shed, be sure to use a high-quality, 100-percent acrylic latex paint. In most cases, you'll also have to apply a primer, unless the siding or trim boards come with a factory-applied coat of primer. Check the label on the can of topcoat paint for specific priming requirements. It's also a good idea to paint the interior walls, which will help stop moisture from passing through the walls from the inside and blistering the paint off the outside. Consider, too, the siding material. If you're using a beautiful wood, such as cedar, redwood, or cypress, do you really want to hide the attractive wood grain behind a coat of paint?

Stain options

Semitransparent stain is an excellent choice for maintaining wood's natural texture. It won't peel or blister the way paint will, but stain will fade over time, especially in areas exposed to direct sunlight. In addition, semitransparent stains are typically available in only wood-tone colors and various shades of gray.

Solid-color—or opaque—stains are formulated to fall somewhere in between paint and

PRO TIP

Can't decide which color to paint or stain your shed? Color photocopies of the building plan with felt-tip markers to create various color schemes.

To give it an aged appearance, this newly built storage barn is sided in rough-sawn pine boards that are finished with a light coat of semi-transparent stain.

A barn-red solid-color stain was used to tint and protect the vertical-board cedar siding on this tool shed with attached firewood-storage extension.

semitransparent stain. They come in a wider variety of colors than semitransparent stains do, but because they contain more "solids," they can peel and blister the way paint does.

A popular shed-finishing technique is to use semitransparent stain on the body of the shed and solid-color stain—in a contrasting color—on the trim.

Clear finishes

Clear wood preservatives require the least amount of maintenance because they don't peel or fade. Although a preservative won't prevent the wood from slowly weathering to a silvery gray color, it will help protect the shed from the harmful effects of sunlight, rain, snow, and mildew. When buying a wood preservative, look for one that contains water repellent and mildewcide.

If you can't decide on a finish, you should at least apply a coat of clear wood preservative to the shed. Exterior wood should never be left unprotected. If you apply a preservative, you'll still be able to paint the shed some time in the future. Just be sure to scrub the siding clean before applying a coat of primer and a topcoat of paint.

Siting the Structure

The next step in the preliminary planning stage is to select an appropriate building site for the shed. This may sound like a simple task—and if you're lucky, it just might be—but if you pick the wrong spot, you'll end up with an unusable outbuilding.

There are several factors that influence site selection, including the size and topography of the property and how you plan to use the shed. Of course, the local authorities will have their say, too. The building inspector must approve both the building plan and the proposed site before he or she will grant a building permit. Here are several important issues to consider as you walk your property in search of the perfect shed site.

Spots to avoid

It's never smart to build a shed at the bottom of a hill or in a low-lying area where water collects. The excessive moisture can rot wood, blister paint, and cause hinges to rust. It'll also promote mold and mildew growth on items stored in the shed. Plus, the ground near the shed will turn into a soggy, muddy quagmire after every rainstorm.

Some of the most attractive sheds I've seen are tucked deep into the woods, completely surrounded by towering trees and lush ground cover. Unfortunately, these lovely woodland settings are ill-suited for storage buildings. First of

This 8-ft. by 10-ft. storage shed was converted into a children's playhouse. It features an arched Dutch door and a giant 30-lite window. Avoid siting a shed in the middle of a wooded lot, where it will prematurely age.

all, the trees and surrounding vegetation block out much of the sun and wind, so the shed receives very little sunshine or ventilation. As a result, the building remains dark and damp, creating the perfect environment for mold and mildew growth.

The shed's roof is especially susceptible and can quickly become overgrown by a thick layer of moss, lichen, mildew, or vines. It's not unusual to find grass, mushrooms, or even small trees growing out of the roof. If you have to mow the roof, you picked the wrong spot for your shed.

Woodland sheds are also under constant assault from falling branches, acorns, leaves, pine needles, and other types of canopy debris. Furry forest creatures are much more likely to move into or under a shed built in the woods. And there's always the potential for severe damage caused by a fallen tree. It takes only one direct hit from a large tree to reduce a shed to a pile of kindling.

Wooded sites require a lot of extra prep work, too. You'll have to cut down trees, pull up ground cover, and scrape the forest floor clean and level. In some cases, you'll even have to dig out boulders and stumps, which is not a fun way to spend a weekend.

TRADE SECRET

If you're building your shed on a slightly sloping plot of land, position the front of the shed (the side with the primary door) on the high side of the slope. That way, the land will slope down toward the rear of the shed and the slanting grade won't be so obvious. Placing the door on the high side will also make it much easier to step into the shed because the door's threshold will be closer to the ground.

ACCORDING TO CODE

In most municipalities there's a rather obscure zoning ordinance on the books, known as a lot-coverage limit. It restricts the total area of land that can be covered by all the buildings on a single lot. The total percentage includes the land covered by the house, the garage, and any outbuildings. When you apply for a building permit, ask whether there's a lot-coverage limit. If there is, make sure the addition of a shed won't exceed the legal limit.

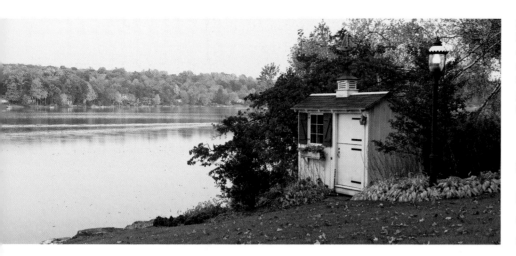

Measuring just 6 ft. by 8 ft., this compact salt-box has board-and-batten siding, three-tab asphalt roof shingles, and a weather-vane cupola. Siting it near trees—not in the woods—takes advantage of a woodland setting but protects the shed from falling branches and mildew growth.

Setback distances

Setback requirements are another important consideration when siting a shed. These are strict town ordinances that establish how far away the shed must be from such things as the following:

• Side, front, and rear property lines
• Streets, driveways, and sidewalks
• Houses, garages, and decks
• Swimming pools and ponds
• Septic tanks and leach fields
• Wetland areas
• Water-well pumps
• Power lines and telephone poles
• Easements

Setback distances vary widely from town to town, even from one neighborhood to the next. They can range anywhere from 10 ft. (leach field) to 100 ft. or more (wetlands). Again, check with the local zoning board or building department for specific information, and be sure to adhere to the letter of the law. If you violate one of the setbacks, the town can legally make you move the shed, which, as you can imagine, is no easy chore.

What if the only decent building site in your entire yard is 10 ft. away from the side property line and the setback requirement is 15 ft.? Are you out of luck? Not necessarily. Most towns let you apply for a variance, which means you have a chance to appear in person before the zoning commission and plead your "hardship." Its members will hear your case and decide whether or not to grant the variance. It can be denied for

SETBACK REQUIREMENTS

Setback distances indicate how far away the shed must be from such things as wetlands, septic tanks, leach fields, property lines, well pumps, and other structures. Check with the local zoning board for the setback requirements in your town. They can range anywhere from 10 ft. or 15 ft. up to 100 ft. or more.

To avoid these problems but still enjoy a lovely woodland setting, build the shed at the edge of the forest. You can even set the building back into the woods a couple of feet, but don't bury it completely.

any number of reasons, including a protest from a neighbor, health and safety concerns, environmental issues, and the ever-changing winds of local politics.

If the variance is granted, you'll be able to build the shed closer than the rule normally allows but—and this is important—no closer than the variance states. In my hypothetical example, the shed could now be 10 ft. from the property line, but not any closer.

The mere thought of having to apply for a variance causes great angst in many homeowners. Don't let this happen to you. Zoning boards are made up of fellow homeowners who are typically very supportive and sympathetic to variance requests. In fact, all the building officials I interviewed said they grant a vast majority of the variances brought before them. One building inspector said that consent is given more than 90 percent of the time.

However, zoning boards often meet only once a month; if the docket is filled for next month's meeting, you'll have to wait until the following month to plead your case. Avoid long construction delays by finding out early on from a zoning board member whether you'll need to apply for a variance.

Easy access

Sheds are often relegated to the deepest corner of the yard, far from the house. And there's nothing wrong with that, if its primary use is to store a riding mower or engage in a specific activity, such as potting plants or woodworking.

However, if it will be an active storage building that your family will use on a regular basis, situate it close to the house or garage. You'll then be able to retrieve an item from the shed without having to traipse across the backyard, a great convenience if you live in an area that gets a lot

of snow or rain. And the kids will be much more likely to put away their bikes, toys, and sports equipment if the shed is close by.

Sunny outlook

To gain the most amount of natural light, position the shed so the wall with the largest windows or widest doorway faces south. If you're using the shed for general storage, its orientation to the sun isn't all that important. But for potting sheds and other specific uses, the sun's solar

SAFETY FIRST

It's seldom necessary to install electricity in a shed that's strictly used for storage, but for safety's sake, consider lighting the way to the shed. Place a floodlight on the house or garage and aim it at the shed. It will prevent you from tripping over or walking into something during nighttime trips to the building.

How are you going to get your building materials to the building site? That isn't a problem if the site is next to your driveway, but what if it's located deep in the backyard, through the fence, past the swimming pool, and beyond the garden? If the lumber delivery truck can't reach the site from your driveway, try a neighbor's driveway or a side street. If that isn't feasible, call your buddies and grab a wheelbarrow; you'll have to haul the lumber piece by piece.

The Phoenix Solar Shed is part storage shed, part greenhouse. The skylights are placed on a south-facing slope to capture the most amount of sunlight. This shed is available as an easy-to-build kit from Handy Home Products.

energy and natural light can help grow plants and illuminate workbenches. To help control the amount of light and solar heat entering the shed, outfit the windows with roller shades or operable wooden shutters.

Code Concerns

It's common knowledge that before you can build anything on your property, including a shed, you must contact the local building department and apply for a building permit. However, few homeowners realize that the building department should be their last stop at city hall when requesting permission to build. That's because the building inspector won't grant a permit until you get approval from the inland wetlands commission, health department, and zoning board. Here's a detailed look at the typical chain of events.

Inland wetlands

Call your town clerk or city hall and ask whether there's an inland wetlands commission. If there is, you must submit a plot plan of your property showing the proposed building site of the shed. If you don't have the plot plan, check the files at the building department. Not there? Then you'll have to hire a surveyor to draw one.

The minimum setback distance from a wetland is usually 100 ft. You may not have wetlands on your property, but you'll need to have that fact verified by the commission. Technically, wetlands are specific types of soil, as designated by the Department of Environmental Protection. Wetlands include obvious areas, such as swamps, lakes, ponds, rivers, and streams. They also encompass places that carry intermittent or seasonal water from any number of sources, including rain runoff from a nearby property or street.

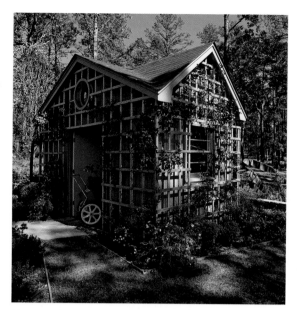

A trellis screen made of crisscrossing 1×2s supports a lush crop of creeping vines. Note that the trellis is mounted 4 in. from the siding. Be aware that you're not allowed to build on property that has been designated as wetlands.

How do you know whether you have wetlands on your property? It should be noted on the original plot plan of the lot, but only if your home was built after the formation of the inland wetlands commission. If your home is more than 15 or 20 years old, it probably predates the commission and the plot plan won't indicate any wetland areas. However, that doesn't mean they don't exist. If necessary, hire a state-licensed soil scientist to check your property for wetlands.

Health department

If your home has a septic system, the health department will want to make sure you don't build the shed too close to the septic tank or leach fields. But more important, it wants to make sure you maintain something known as a "viable reserve area."

Septic systems sometimes fail, and when they do, they must be replaced. The town will want

PRO TIP

Don't allow the lumber delivery truck to drive across your lawn during or immediately after a rainstorm. It'll tear up the grass and severely compact the wet soil.

PRO TIP

Before going to the building inspector for a permit, first get approval from the inland wetlands commission, health department, and zoning board.

to make sure there's enough open land on your property to install a new, often bigger septic system should your current one fail. This reserve area must also be large enough to permit heavy equipment to come in and install the new tank and leach fields. Therefore, the shed can't be built near the existing septic system or within the viable reserve area.

If you don't have a septic system but have city water and sewer instead, you have a little more flexibility regarding where to build the shed. However, you still must get the municipal water company to sign off on your plot plan, so be sure to call the water company and ask about any building-site restrictions.

Zoning board

The inland wetlands commission and health department tell you where you can't build. Go to the zoning board to find out where you can put the shed.

As mentioned earlier, local zoning boards establish setback distances and handle variances. As long as the shed site doesn't violate any of the setbacks, there's no reason for the board to deny your request to build. If necessary, you can always apply for a variance.

Building department

After you've run the gauntlet and received approval from the three offices listed previously, bring your shed plans to the building department. The building inspector's job is to make sure the shed conforms to all local and national building codes. He or she will carefully check the foundation type, the grade of plywood specified, and the size and spacing of the lumber used for the floor joists, wall studs, and roof rafters. Be sure to read Chapter 2 for specific information on various shed-building techniques.

The 6-ft.-wide roof extension on this shed is sheathed in lattice so that dappled sunlight can reach the potting benches below.

The most critical building code concerns the type of foundation required. There will be more on this subject in the next chapter, but be aware that there are two basic types: on-grade and permanent foundations. The code typically permits small to medium-size outbuildings to be supported by an on-grade foundation made up of concrete blocks or weather-resistant wooden timbers set directly on the ground. (The sheds shown in Chapters 4, 5, 6, and 7 all feature on-grade foundations.)

However, bigger sheds—usually those larger than 200 sq. ft. or taller than 12 ft.—require a permanent or frost-proof foundation (see the Gambrel Storage Barn in Chapter 8). For this type, you must dig down past the point where the ground freezes in winter, known as the frost line, and then pour concrete footings or piers. The frost-line depth varies, depending on your region of the country. In cold climates where problems associated with frost heave are most common, the frost-line depth ranges from 36 in. to 48 in. Ask the building inspector for the exact depth in your area.

PRO TIP

When hiring a pro, be sure to get a written estimate stating the extent of work to be done, the timeline for doing it, and the cost for labor and materials.

WHAT CAN GO WRONG

When a professional contractor fails to show up on the day promised, it can throw off your construction schedule and spoil an entire weekend that you had planned to work on the shed. Avoid this problem by hiring the contractor as early as possible. Call at least a month ahead of time, and get him or her to commit to a date. Then call again one week before the scheduled date to confirm the time of arrival. Remember, it's the squeaky wheel that gets the contractor.

Every town charges a building-permit application fee. Sometimes it's a standard amount of $15 to $25, but often the fee is based on the overall cost of the outbuilding. For example, the fee might be $5 per $1,000 of cost. Therefore, the application fee for a $3,000 shed would be $15.

On-site inspections

When you finally receive the building permit, ask whether any inspections are necessary. Most towns require inspections for permanent foundations and for adding electricity to a shed. The building inspector will want to see the footing holes before you fill them with concrete to verify that they're dug to the correct depth.

If you're running an underground electrical cable from the house to the shed, the inspector will want to see the trench before you backfill it to confirm that it's deep enough and that you're using the appropriate cable. Inspections are also required for the rough wiring that feeds the wall receptacles (outlets), light fixtures, and switches inside the shed.

Even if your shed doesn't have a permanent foundation or electricity, chances are good that you'll still need to have a final inspection. The building inspector will need to see the completed shed to verify that its size, location, and construction are in accordance with the approved building permit.

Hiring Help

We all need a little professional help from time to time. As I mentioned earlier, the goal of this book is to provide you with the necessary information and confidence to build your own shed. However, that doesn't mean you can't hire a professional contractor for certain aspects of the project.

There's no shame in hiring a pro to tackle jobs that you don't have the time or inclination to do or that you simply can't do because you don't possess the skills, experience, or special equipment. Here are my top three reasons for calling in a contractor:

The big dig

Hire a pro whenever a site requires extensive excavation. A single backhoe can clear and level the average shed site in less than an hour. Using a shovel and rake may take you two or three weekends.

When it comes to digging footing holes for a permanent foundation, you have a couple of options. If the soil is soft and easy to dig, you can save yourself some cash by using a manual post-hole digger or renting a gas-powered auger. However, if there's even the slightest chance that you'll run into rock or clay, hire an excavator.

SAFETY FIRST

It's often necessary to rent a specialized tool, such as a power auger, chainsaw, concrete mixer, or stump grinder, to tackle a particularly difficult or time-consuming chore. However, using a powerful piece of equipment with which you're unfamiliar can cause serious injury. Make sure the tool-rental dealer shows you how to safely operate the equipment and explains any particularly dangerous characteristics.

Get two sheds in one with this cleverly designed 6-ft. by 16-ft. building from Walpole Woodworkers. On the left is a workshop; to the right, behind the double doors, is general storage. When electrifying an outbuilding, play it safe and hire a licensed electrician.

TRADE SECRET

Mount an exterior light fixture to the shed, and install three-way switches at the house and shed. The light can then be turned on and off from either location.

Power Play

ONE WAY TO BRING electricity to a shed—particularly if it is located in a relatively remote area—is by installing a solar-powered panel. The panel contains several photovoltaic cells that capture the sun's energy, which is then stored in a rechargeable battery. Keep in mind that a small solar panel like the one shown here can produce only enough power for a couple of lighting fixtures. You won't be able to run machinery.

Concrete solutions

Mixing and pouring concrete for a few footings or piers isn't all that much work. In fact, it's kind of fun if you've never done it before. In most cases, you won't even need to rent a portable mixer; a wheelbarrow and a hoe will work fine.

However, if the shed has a concrete-slab foundation and you're new to concrete work, you may want to hire a mason. Pouring a slab isn't that difficult, but if it's done incorrectly, it's nearly impossible to fix once the concrete has cured. Plus, it requires a few specialty tools that most homeowners don't own, such as a bull float, a hand edger, and a finishing trowel.

For an additional fee, a mason will also build the wooden form and prepare the site for the pour, but any competent do-it-yourselfer can easily complete those tasks.

Getting wired

Depending on how you plan to use your outbuilding, you may or may not need to wire it with electricity. If it'll be used for general storage, there's really no need to electrify the building. You can use a flashlight for the few times you may need to enter the shed at night. However, if you plan to use the shed as a woodshop, home office, or arts and crafts center, you'll obviously need electricity for lighting, power tools, computers, and other electronic equipment.

Ordinarily, power is brought to a shed from the house via an underground cable, which is buried in a trench and hooked up to an existing house circuit. Occasionally, it's easier to run a new circuit from a nearby utility pole. Except for basic tasks, such as installing electrical boxes or hooking up switches and outlet boxes, making electrical connections is work that should be done by a licensed electrician.

You may be able to save the electrician some time—and yourself some money—by doing some of the preliminary work. For example, if the electrician plans to pull power from the house, you can dig the trench and run the cable to the shed. Be sure to dig down to the code-specified depth (usually 18 in.) and use only an underground feed (UF) cable, which is specifically designed for burial. Then have the electrician hook up the cable to the main panel in the house and to the receptacles, switches, and other devices in the shed. Don't forget that the building inspector will want to see the cable and trench before it's backfilled, and inspect all the rough wiring in the shed.

DESIGN
OPTIONS
STORAGE SOLUTIONS

▲ A pegboard panel attached high on an interior wall provides a convenient way to hang lawn-and-garden hand tools. Be sure to use ¼-in.-thick pegboard; the ⅛-in. type is too thin and weak.

▶ It only takes a few 2×4s and some ¾-in.-thick plywood to create a simple storage bench. The one shown here measures 24 in. wide by 48 in. long, and is used for potting plants. The high backsplash keeps soil from falling behind the bench.

◀ An overhead storage loft is ideal for safely storing out-of-season items, such as lawn-chair cushions, snow skis, and swimming pool supplies.

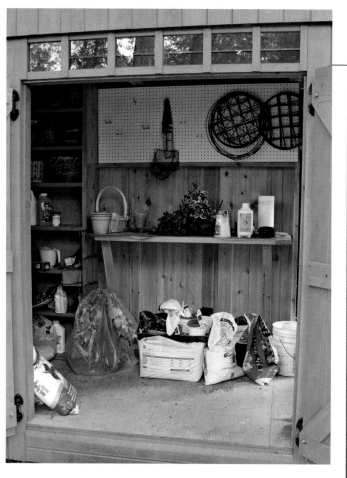

◀ In this shed, a partition wall was built 4 ft. inside of the doors to create a small garden-storage area. The larger, main section of the shed is accessible through a second set of doors on the front façade.

▼ An 8-ft. by 10-ft. extension built onto the side of a shed provides the perfect place to store firewood, bicycles, lawn mowers, and other items that don't easily fit into the shed.

▲ When looking for extra storage, don't overlook the space on the back side of the doors. Here, a wire storage rack was fastened to the door, providing seven roomy shelves.

CONSTRUCTION METHODS

There are many ways to build a storage shed. In this chapter, I'll explore all the possibilities for each construction step, from creating a stable foundation to framing the roof. In many instances, you'll have a choice between three or four building techniques. For example, you can make the shed's floor out of plywood, brick, concrete, or gravel. The walls can be framed with 2×4 studs or 4×4 posts and beams. At the door of the shed, you can build steps, a ramp, or an entrance deck.

The specific techniques that you choose will often be based on personal preference, degree of difficulty, topography of the building site, and—of course—cost. However, be aware that you won't always be the one to choose the construction method. In certain cases, how you build your shed will be determined by the local building inspector. For example, if the shed is larger than a certain size, you'll most likely be required to build a frost-proof foundation. The building department will also dictate the dimensions and spacing of the framing lumber used. ▶ ▶ ▶

Fiber-form tubes are commonly used to build permanent, frost-proof foundations. Each tube is set in a hole dug below the frost line, then filled with concrete.

Solid-concrete blocks are ideal for building on-grade foundations. The blocks are aligned in rows and checked for level with a long, straight 2×4 and a 4-ft. level.

Choosing a Foundation

The success or failure of any outbuilding relies heavily on its foundation. No structure—regardless of how well it's designed or built—will survive for very long on a weak or poorly made base. Therefore, choosing and building a proper foundation is the single most important construction step in the entire project.

Shed foundations fall into two basic categories: on-grade and frost-proof. On-grade foundations (sometimes called "floating foundations") sit right on the ground and are sufficient for all but the very largest outbuildings. They're also the quickest and simplest to build because they don't require you to dig deep holes or pour concrete footings or piers. On-grade foundations are usually made of pressure-treated lumber (see Lean-to Shed Locker on p. 76) or solid-concrete blocks (see Saltbox Potting Shed on p. 124).

Permanent, frost-proof foundations are more difficult to build, but they're by far the strongest and longest lasting. These types of foundations are designed for cold-weather regions where ground movement caused by freeze/thaw cycles can affect a building.

The most popular material for building frost-proof foundations is poured concrete, which can take the form of a footing, pier, or slab. You can also create a permanent foundation—and avoid frost-heave troubles—with an age-old building technique called pole-barn construction. In this system, several tall round poles or square posts set into deep holes support the structure (see Gambrel Storage Barn on p. 186).

The best foundation to build for your shed will largely depend on what the building inspector recommends, but keep in mind that it's often based on three key factors: the shed's size, the

Pole-barn frames consist of round poles or square posts set in deep holes. Here, a pressure-treated 4×4 post is used.

What Is Frost Heave?

FROST HEAVE SOUNDS LIKE SOMETHING that happens after you eat too many snow cones. But if you live in a cold-weather region, it's no joking matter. Frost heave is a phenomenon that occurs when moisture in the soil freezes and then expands, pushing the ground—and anything buried in it—upward. When the ground thaws, the soil slumps back down. As you can imagine, all this up and down movement isn't very good for a shed. Frost heave can raise a building 3 in. or 4 in. off the ground, then drop it back down. The problem is, the building seldom settles back into its original square and level position.

Fine-grained sand and stiff clay soils are more susceptible to frost heave than coarse-grained soils and compacted gravel are, but the phenomenon can occur virtually anywhere, especially if the building site doesn't effectively drain away groundwater.

The best way to prevent this unsettling condition is to dig past the point where the ground freezes—an area known as the frost line—and then pour a concrete pier or footing. How deep you need to dig depends on how far down the frost penetrates the soil in your particular corner of the world. For example, in New England and the Upper Plains, you may need to dig 42 in. or more. In the mid-Atlantic region, 24 in. to 36 in. may be sufficient. Check with the local building department for the exact frost-line depth in your town.

Frost Heave
Frost heave occurs when moisture in the ground freezes and expands, then thaws and settles unevenly. To avoid frost heave, dig footing holes below the frost line.

Wood post

Frost heave

Frost heave

Rubble falls into hole under post

Frost line

region of the country in which you live, and the type of shed floor you desire. To help you choose the best foundation for your shed, let's take a close look at seven foundation systems: four on-grade types and three frost-proof ones.

On-Grade Foundations

It's no surprise that most sheds are designed to be built with an on-grade foundation. This base is quick and easy to build, relatively inexpensive, and adaptable enough to accommodate all but

the most severely sloping sites. In addition, the components are small and light enough to easily set into place and shift around, making it very easy to get everything square and level.

Solid-concrete blocks

In this type of foundation, the shed is supported by a series of solid-concrete blocks that are laid out in straight, evenly spaced rows. The number of blocks needed and the spacing between them is determined by the size of the shed and the lumber used for the floor joists. For example, the

Eight solid-concrete blocks are arranged in two rows to form this on-grade foundation. Identical diagonal measurements indicate a square layout.

TRADE SECRET

When building a solid-concrete block foundation, it's important that all the blocks be level. However, it's equally important that the blocks in each row be perfectly aligned. The best—and fastest—way to line up the blocks is with a taut string. Install the first and last block in each row. Then stretch a length of mason's line along the edge of the two end blocks and use it as a guide to set the intermediate blocks.

In most cases, you'll need both 2-in.-thick patio blocks and 4-in-thick solid-concrete blocks to build an on-grade foundation.

PROTIP

Gravel and crushed stone are available in 50-lb. bags, but it's cheaper if you buy it by the truckload from a garden shop, nursery, or masonry supplier.

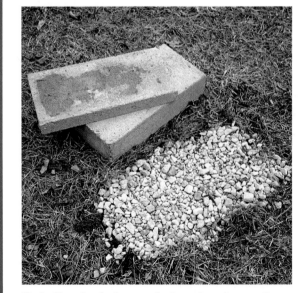

A shallow bed of gravel placed underneath concrete foundation blocks aids drainage and helps prevent them from sinking into the soil.

8-ft. by 12-ft. Saltbox Potting Shed (see p. 124) has 2×6 floor joists. Its foundation is made up of eight blocks set in two rows. A 10-ft. by 16-ft. shed requires 12 blocks arranged in three rows.

It's important to note that you must use only solid-concrete blocks for this type of foundation. Standard wall block or any other hollow block will eventually crack and crumble under the weight of the shed.

The blocks measure 8 in. wide by 16 in. long and come in 4-in.- and 2-in.-thick units. The thicker blocks are placed first, with the thinner "patio" blocks laid on top when you need to raise one block even with the others. In some cases, you may need to stack two or three 4-in. blocks on top of each other to raise the lowest corner of the foundation so it is even with the highest corner.

If the building site is high and dry, you can set the blocks directly on the ground. However, if there's any chance that rain runoff will occasionally drain under the shed, you'll need to use a shovel to remove a patch of grass under each block, compact the soil with a hand tamper, then cover the exposed dirt with 2 in. or 3 in. of gravel before setting the blocks.

Precast pier blocks

This building method is similar to the solid-concrete block foundation. However, instead of using flat blocks, a series of precast concrete pier blocks are used to support the shed's floor frame. The pyramid-shaped blocks are designed for building decks, but they work great for sheds, too—provided you choose the right type.

There are a few styles of pier blocks available, including one that has a square hole molded into the top into which a vertical 4×4 post can be inserted. Another type has a flat wood block set into the top so you can toenail a joist in place. For building shed foundations, I prefer to use Dek-Block® piers. Each block measures 8 in. high by 11 in. sq. and weighs about 45 lb. Molded into the top surface are a 3½-in.-sq. recessed socket and a pair of 1½-in.-wide slots. The socket accepts a 4×4 post; the slots are used to support a 2× floor joist. Because Dek-Block piers can accept either a joist or a post, they can be used on very uneven sites and badly sloping terrain.

A series of precast pier blocks, arranged in three straight rows, provides a simple, secure way to support a floor frame.

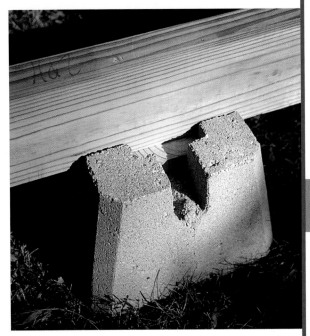

A 2× joist fits into a slot molded in the top of this Dek-Block pier; the concrete pier will also accept a vertical 4×4 post.

PRO TIP

When mixing concrete in cold weather, use warm water to speed the curing process. If it's hot outside, mix in cold water to slow it down.

TRADE SECRET

For maximum strength, dig pier holes about three times wider than the post or concrete block they will support. That's usually 12 in. to 18 in. in diameter.

LEVELING A SLOPED SITE

Straight 2×4

4-ft. level

Asphalt shingle shim

2-in. patio block

4-in. blocks

Gravel

Gravel

After laying out all the blocks for the shed foundation, use a long, straight 2×4 and a 4-ft. level to make sure each block is level with the others. If you need to raise a block just a little bit, place a thin shim of wood or asphalt shingle on top.

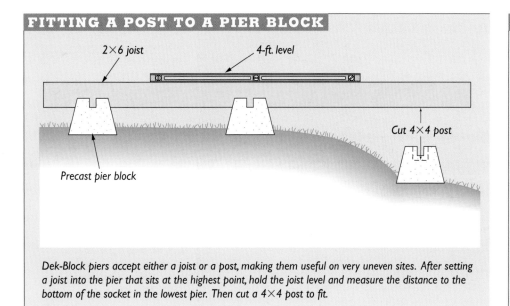

Dek-Block piers accept either a joist or a post, making them useful on very uneven sites. After setting a joist into the pier that sits at the highest point, hold the joist level and measure the distance to the bottom of the socket in the lowest pier. Then cut a 4×4 post to fit.

SKID FOUNDATION

A skid foundation is a simple and effective way to support the floor frame of a small shed.

PRO TIP

Don't bury foundation blocks or skids too deep into the ground. There must be at least 4 in. of airspace under the shed.

Skid foundations

When it comes to time-tested building methods, it's hard to beat a skid foundation. Builders have been using this type of on-grade foundation to support outbuildings for more than three centuries. The technique is surprisingly simple in both concept and application: Two or more long, straight timbers (skids) are laid on the ground in parallel, evenly spaced positions. The building's floor frame is then built on the skids, which are sometimes called runners or deadmen.

Skid foundations are still popular today, and it's easy to see why: They're very fast and easy to build, and they distribute the building's weight evenly over a broad surface. Unfortunately, because the timbers are long and straight, this type of foundation is suitable only for sites that are relatively flat.

Originally, skids were nothing more than logs placed on the ground. Today, they're usually made of pressure-treated 4×6s, 6×6s, or 8×8s. You can also make skids by gang-nailing together three or four 2×6s or 2×8s and setting them on edge.

Although skids are often set directly on the ground, I prefer to lay them on a bed of gravel. The stone creates a very stable base that's not likely to settle or wash away. Begin by laying the skids in position on the ground, then mark around each one using spray paint or flour sprinkled from a can. Move the skids out of the way, then use a flat shovel to remove the sod and about 2 in. of soil from the marked areas. Check the excavated areas to make sure they're close to being level. If they're not, remove a little more soil from the high spots. Next, add 3 in. to 4 in. of gravel. Compact the gravel with a hand tamper or gas-powered plate compactor, then replace the skids.

Timber-frame foundations

This foundation consists of little more than a rectangular wooden frame sitting on a gravel bed. The shed walls are built on the frame, and the entire weight of the building is transferred directly to the ground. However, the real advantage with this type of foundation is that you get to choose from a variety of flooring options. For

SKID FOUNDATION ON A SLOPE: TWO METHODS

If the site is out of level by more than 4 in. or so, it's easier to block up the low end of the skid than it is to excavate beneath the high end. The two best ways to do that are to make up the difference with solid-concrete blocks (A) or stack two skids on top of each other at the low end (B).

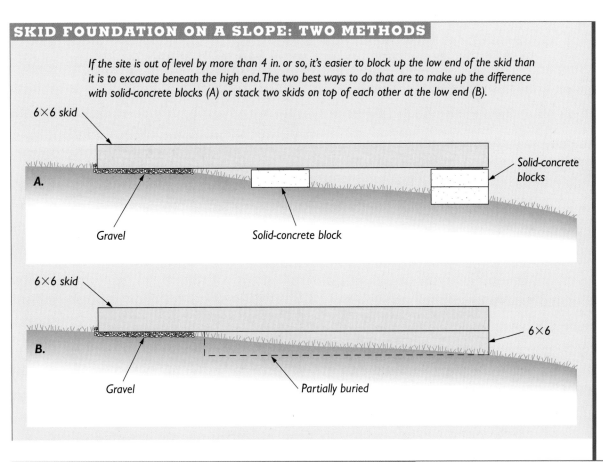

6×6 skid

A.

Gravel

Solid-concrete block

Solid-concrete blocks

6×6 skid

B.

Gravel

Partially buried

6×6

TIMBER FRAME FOUNDATIONS: TWO OPTIONS

A sturdy timber-frame foundation can be built with either half-lap joints (A) or stacked corners (B); it is useful when a gravel, stone, or brick floor is needed for the shed.

A) Half-lap Joint Construction

Two landscaping spikes or screws per corner

Half-lap corner joint

Half-lap corner joint

B) Stacked Corner Construction

Stacked corner joint

Landscaping spikes or screws spaced 24 in. apart

What Is Pressure-Treated Wood?

MORE THAN SIX BILLION board feet of pressure-treated (PT) lumber is sold annually, yet most people know very little about this ubiquitous green-tinted wood. Here are the facts.

Pressure-treated lumber is an exterior-grade wood that's chemically treated to resist rot, decay, and termites and other wood-boring insects. It's milled from softwood trees (typically southern yellow pine) and used for decks, fences, sheds, picnic tables, and playground equipment.

Waterborne chemical preservatives are pumped deep into the wood's fiber under extremely high pressure. Field tests conducted by the U.S. Forest Service have shown that treated lumber can resist decay and termite attack for more than 40 years. However, not all treated wood is created equal. The level of rot resistance is directly related to the amount of chemical preservatives in the wood. Lumber that's stamped "Above Ground Use" should be used only where it won't touch the ground, such as a deck railing or a fence board. Lumber that's designated "Ground Contact" can be used on or in the ground.

To make sure you're building with the right stuff, check the lumber's label or stamp for its "Chemical Retention Level." This number represents the minimum amount of preservative retained in the wood and is expressed in pounds of preservative per cubic foot of wood. The higher the number, the more rot resistant the wood is. The American Wood Protection Association has established the following retention levels for lumber treated with chromated copper arsenate:

Lumber with a .40 retention level is adequate for building shed foundations, but for additional protection—and a little extra peace of mind—upgrade to a .60 retention level when building a pole-barn, timber-frame, or skid foundation.

For more than 60 years, the chemical most commonly used in treated lumber was chromated copper arsenate (CCA). However, due to health concerns, CCA-treated lumber for residential use was banned back in 2003. (It's still used for telephone poles, docks, boardwalks, and other large-scale commercial projects.) The pressure-treated lumber widely available today at lumberyards and home centers is treated with an inorganic chemical (instead of arsenate), and it's typically referred to as ACQ lumber, which stands for the chemicals alkaline copper quaternary. This new-generation outdoor lumber is sold under various trade names, including Wolmanized® Outdoor Wood, Preserve, Thompsonized® Wood, and NatureWood®.

Treated lumber is labeled with information that includes chemical-retention levels and recommended uses. For foundations, use only lumber that's rated "Ground Contact."

Chemical Treatment

Retention Level (lb./cu.ft.)	Retention Level (lb./cu.ft.)
.25	Above Ground
.40	Ground Contact
.60	Wood Foundation
2.5	In Salt Water

example, I used brick pavers to create a handsome, hard-wearing floor for the Lean-to Shed Locker on p. 76. The floor area within the timber frame could also be filled with gravel, concrete, crushed granite, marble chips, or slabs of bluestone or slate. Another flooring option is to nail pressure-treated 2×6s over the frame in a manner similar to that of a deck. Just keep in mind that the 2×6s must be installed before you erect the shed walls.

A foundation frame is typically made from pressure-treated 4×4s, 4×6s, or 6×6s. The timbers are joined with half-lap corner joints or stacked two or three high and fastened together with long landscaping spikes or screws.

Frost-Proof Foundations

Frost-proof foundations extend deep into the ground to prevent freeze/thaw cycles from upsetting the building. They're generally required

Calculating Concrete

TO FIGURE OUT HOW MUCH CONCRETE you'll need for pouring a slab or footing, refer to the chart at right or use the following formula: Multiply the length times the width times the thickness of the slab, then divide by 12. Divide again by .45 for the number of 60-lb. bags of ready-mix concrete you'll need. If you're using 80-lb. bags, divide by .60 for the number of bags needed.

To estimate how much concrete you'll need for the piers, use the chart at below. Note that there's information for both 60-lb. and 80-lb. concrete bags and for holes ranging in depth from 24 in. to 42 in. and in diameter from 8 in. to 18 in.

Concrete Slab

4-in.-Thick Slab (total sq. ft.)	Number of 60-lb. Bags of Concrete Needed	Number of 80-lb. Bags of Concrete Needed
6 ft. by 8 ft. (48 sq. ft.)	36	27
8 ft. by 10 ft. (80 sq. ft.)	59	44
10 ft. by 12 ft. (120 sq. ft.)	89	67
12 ft. by 14 ft. (168 sq. ft.)	124	93

Number of Bags of Concrete for Piers

Depth of Hole	60-lb. Bags				80-lb. Bags			
	Diameter of Hole				Diameter of Hole			
	8 in.	10 in.	12 in.	18 in.	8 in.	10 in.	12 in.	18 in.
24 in.	1.5	2.5	3.5	8.0	1.25	2.0	2.66	6.0
30 in.	2.0	3.0	4.5	10.0	1.5	2.5	3.33	7.5
36 in.	2.33	3.6	5.25	12.0	1.75	3.0	4.0	9.0
42 in.	2.75	4.25	6.25	13.75	2.0	3.25	4.6	10.5

Generally speaking, you're not allowed to pour concrete for a frost-proof foundation until the building inspector has examined the holes. This is the building department's only way to know whether the holes are dug to the proper depth. Note that most inspectors are obliged to respond to your call for an inspection within a certain time period, usually 24 hours.

by code in cold-weather regions for sheds larger than 200 sq. ft. or taller than 12 ft. However, building codes differ from state to state, so check with the building department for the exact requirements in your area.

Frost-proof foundations are typically more difficult to build than on-grade types, but how much harder depends on the soil. If the ground is soft

POURED-CONCRETE PIERS: THREE METHODS

There are many ways to pour a pier foundation. Here are three techniques that work well for virtually any size outbuilding:

Solid Pier with Pier Block
Dig 12-in.-dia. holes down to the frost line, then pour concrete piers flush with the ground. Set precast concrete pier blocks or solid-concrete blocks on top. The blocks can be set in the wet concrete or simply laid on top once it cures. If you set them in the wet concrete, make sure all the blocks are level before the concrete hardens.

Sonotube Pier
Pour a 6-in. to 12-in. pad of concrete—called a footing—into the bottom of the hole and let it cure overnight. Next, stand a round fiber-form tube (commonly known by the trade name Sonotube®) on top of the footing. Backfill around the tube with soil, then fill it with concrete. The advantage of this method is that before installing the tube, you can cut it flush with the ground or let it protrude above the hole to create a raised pier.

Formed Pier
Build a square form out of 2× lumber and set it over the center of the hole. Hold the form in place with a couple of 1×3 stakes pounded into the ground. Then fill the hole with concrete, bringing it right to the top of the form. After the concrete cures, strip away the form to reveal a raised pier.

Plan View of Form

and sandy, you may be able to excavate it with a shovel and post-hole digger; if it's hard-packed clay or very rocky, bring in a backhoe.

Poured-concrete piers

In its simplest form, a pier is nothing more than a column of concrete poured into a hole that extends below the frost line. Two or more rows of piers are used to support the shed's floor frame, similar to the way solid-concrete blocks are aligned for an on-grade foundation.

There is a wide variety of galvanized-metal framing hardware—such as post anchors, beam connectors, and tie-down straps—that can be used with poured-concrete piers. These specialty brackets provide a quick, easy way to create strong, lasting joints between the concrete piers and the wooden joists, posts, or carrying beams. Just make sure you set the brackets in the piers before the concrete hardens.

Poured-concrete slab

The average backyard storage shed doesn't really need a poured-concrete floor, but it's the best choice for large outbuildings that will be used to store heavy equipment, such as woodworking machines, tractors, boats, motorcycles, snow-mobiles, and antique cars.

There are two basic ways to pour a concrete floor, but only one qualifies as a frost-proof foundation. It's called a monolithic slab, because the floor and the perimeter foundation walls are all poured at the same time. The walls extend down to the frost line and are usually between 12 in. and 16 in. thick. The floor itself is only 4 in. to 6 in. thick, but it's reinforced with wire mesh or metal reinforcing bars.

The second type of concrete floor is known as a floating slab or an on-grade slab. It's nothing more than a 4-in. to 6-in. layer of concrete sitting on the ground. This type of floor should never be used when the plans or local building codes call for a frost-proof foundation.

Pole-barn foundations

All the foundations discussed so far are designed to support the floor of an outbuilding. The floor, in turn, supports the walls. A pole barn is completely different. In fact, it technically doesn't even have a floor.

Pole-barn construction starts with a series of holes dug below the frost line around the perimeter of the foundation. Concrete footings are poured into the bottom of each hole, then tall, decay-resistant round poles or square timbers, which extend all the way to the tops of the walls, are set in the holes. Horizontal beams are bolted along the tops and bottoms of the poles to tie everything together and support the walls and roof framing.

TRADE SECRET

Hardware can be attached to cured concrete with specialty masonry fasteners, which can make layout a little easier during a messy concrete pour. It's always best to avoid drilling into concrete, and screw-down brackets aren't nearly as strong as cast-in-place ones are. You'll find a variety of systems for fastening hardware to concrete at your local hardware or building supply store. I've found that epoxy-encapsulated fasteners work best.

Is Burying Posts a Good Idea?

THERE'S SOME DEBATE ABOUT whether or not you should bury a wood pole—even a treated one—in concrete, where it may eventually rot. Some builders prefer to pour raised concrete piers and set the poles on top (see the drawing on p. 36). This method does help the poles last longer, but much of the structural integrity of a pole-barn foundation comes from the fact that the poles extend deep into the ground. Raising them out of the ground can weaken the structure.

Ordinarily, I don't like to bury wood in the ground, but it's definitely the best way to gain the strength needed when building a pole barn or setting a gate post or an end-of-the-run fence post. I've buried many pressure-treated posts over the years and have always questioned the process, but on the other hand, I've never had to replace any of them. I spoke with a local pole-barn builder who has been burying wood—pressure-treated and creosote-soaked—for more than 30 years and continues the practice to this day with confidence.

WHAT CAN GO WRONG

It doesn't seem possible that there's a wrong way to dig a hole, but there is when you're digging a hole for a concrete footing. First, don't make the hole any wider than necessary; you don't want to disturb the surrounding soil. More important, make the hole wider at the bottom than at the top to create a broad, stable base.

PRO TIP

When working with large timbers, swap your 16-oz. hammer for a 22-oz. to 26-oz. framing hammer, which has a longer handle and a heavier head.

Post in Sonotube
This basic pole-barn foundation is strong and simple to build; the poured concrete footing creates a solid, frost-proof base, while the tube and stone mixture leave little for frost to grab onto.

The diameter of the hole is three times wider than that of the post.

Post on Footing
A simpler option is to eliminate the tube and pour the concrete footing directly into the bottom of the hole. Then after placing the vertical post on top of the footing and centered in the hole, backfill the entire hole with a mix of crushed stone and sand.

The hole is dug wider at bottom.

Post on Raised Pier
Raising the pier above ground level by pouring a concrete pier avoids the problem of rot, but it may weaken the strength of a pole-barn foundation.

Instead of a wood-framed floor, a pole barn's floor is actually the ground. To keep it from wearing away or turning muddy, the area is covered with several inches of processed stone, pea gravel, or wood chips. As a result, the floor is basically flush with the surrounding grade, making a pole-barn foundation perfect for out-buildings that house lawn tractors, boat trailers, farm machinery, horses, and livestock.

As with other types of foundations, there are several ways to build a pole barn. For the Gambrel Storage Barn (see p. 186), fiber-form tubes were placed in the holes, then an 8-in.-thick concrete footing was poured into the bottom of each tube. After the concrete cured, 4×4 pressure-treated posts were placed on top of the footings and another 12 in. or so of concrete was poured in. The rest of the tube was then filled to the top with a mixture of crushed stone and sand.

Wall-Framing Techniques

There are three common construction techniques used to build shed walls. A vast majority of outbuildings—probably more than 90 percent—are stick-built out of 2×4s. To be honest, it's hard to justify using any other technique, especially if you're building a basic storage shed. However, if you're building a work-shop, writer's studio, or home office, you may want to put up a post-and-beam building. The exposed timbers lend a more finished, hand-crafted look to the interior. This is also a more interesting and challenging way to build.

A pole-barn building, as discussed previously, is suspended off the ground by several round poles or square posts. The wall framing is fastened to the poles and can be made out of large tim-bers or 2× lumber. The exact wall-framing technique depends on the type of poles or posts used and your choice of siding material.

Stick-built construction basics

It's no mystery why carpenters prefer stick-built construction: It's the fastest, easiest, and most affordable way to frame walls. The term stick-built refers to the fact that the walls are constructed out of individual "sticks" of lumber: typically 2×4s or 2×6s. This technique is also called western or platform framing.

Headers are an important component in stick-built construction and are used to transfer roof loads above doors and windows down through the framing and onto the foundation. Their size and thickness depend on several factors, including the width of the opening, the height of the building, and whether the wall is a load-bearing one.

You can use standard framing lumber for building walls; Douglas fir, hemlock, and spruce are fine. However, it's important to use pressure-treated lumber for any framing member, such as the mudsill, that comes in contact with the ground or a concrete foundation.

Shed walls are typically framed with 2×4s. Save time by building the walls on the ground, then tip them up into place.

WHAT CAN GO WRONG

One of the most common and chronic of all carpentry mistakes occurs when framing a rough opening in a wall for a window or door. The problem is caused by the somewhat confusing way that manufacturers and retailers specify the size of a window or door. For example, a 48-in.-wide window is sold as a 4-0 unit (pronounced "four-oh"). A 3-8 door isn't 38 in. wide; it's 44 in. wide. If you're not paying close attention, it's very easy to frame a 30-in.-wide opening for a 3-0 door, then discover that the opening is 6 in. too narrow. Avoid these costly, time-consuming mistakes by double-checking the dimensions on the plans before ordering windows and doors. Also, place your order early and have the units on hand before starting construction. That way, you can measure each window and door before framing the rough openings.

STICK-BUILT WALL

Top plate

Jack or trimmer stud

Cripple

Rough sill

Double header

Full studs

Cripple stud

Bottom plate

Jack or trimmer stud

King stud

Each section of a stick-built wall consists of horizontal top and bottom wall plates and vertical studs, which are usually spaced 16 in. on center. Horizontal beams, called headers, run across the top of each window and door opening and transfer weight from above the window or door to the jack and king studs on each side of the opening.

SAFETY FIRST

Pressure-treated wood poses no health hazards if you follow a few simple precautions. Be sure to wear gloves when handling treated lumber, and wash up thoroughly before eating or drinking. Always wear safety goggles and a dust mask when cutting or drilling treated wood, and be sure to properly dispose of all scraps; never burn treated wood.

The frame of a post-and-beam building has far fewer parts than does a stick-built shed, but the parts are much heavier.

Post-and-beam construction

This ancient building method uses large vertical posts and horizontal beams to form the skeletal frame of the walls. There are far fewer parts in a post-and-beam wall compared to those in a stick-built wall, but the parts are larger, so they're a little more difficult to work with. The posts and beams are usually cut from 4×6s, 6×6s, or 8×8s.

There are various ways to join the posts to the beams, but the traditional method of using mortise-and-tenon joinery is still preferred today. Each joint consists of a recessed slot (mortise) and a protruding tab (tenon). Mortises are cut by first drilling out the waste wood with a large-diameter drill bit, then squaring up the hole with a mallet and chisel. Tenons are typically cut with a portable circular saw and a handsaw.

A post-and-beam structure can be covered with plywood sheathing and siding like any other building. Or the siding can be set within the framework of the building so the posts and beams remain visible from both inside and out (see the right photo on p. 10).

Nominal Numbers

FRAMING LUMBER IS REFERRED TO by its "nominal" dimension, not by its actual dimension. That's because when a log is ripped into 2×4s during its initial pass through the sawmill, the roughsawn boards measure a full 2 in. by 4 in. However, after the final milling stage and drying process, they're only 1½ in. thick by 3½ in. wide. In fact, all 2× lumber is actually 1½ in. thick, and all 1× boards are actually ¾ in. thick. Use the chart below as a handy reference for both nominal and actual dimensions of standard lumber sizes.

Nominal Dimension	Actual Dimension	Nominal Dimension	Actual Dimension
2 x 3	1½ by 2½	2 x 12	1½ by 11¼
2 x 4	1½ by 3½	4 x 4	3½ by 3½
2 x 6	1½ by 5½	4 x 6	3½ by 5½
2 x 8	1½ by 7½	6 x 6	5½ by 5½
2 x 10	1½ by 9½	8 x 8	7¼ by 7¼

PROTIP

A power miter saw won't replace your portable circular saw, but it will offer a quicker, more accurate way to crosscut framing lumber.

◄ **Fit the beam's tenon into the mortise in the post. Hold the joint closed with long wooden dowel pegs or galvanized screws.**

Pole-barn framing

The method used to frame the walls of a pole barn depends largely on whether it has round poles or square posts. For round poles, it's easiest to nail 2×4s or 2×6s horizontally across them. If the walls will be covered with plywood siding, three boards should suffice. If you're planning to install board-and-batten siding or some other vertical-board exterior, you'll need to install four or more equally spaced 2× boards to provide adequate nailing for the siding.

If the building has square posts, you have two options: Nail 2×4s or 2×6s across the posts as described for round poles, or build framed sections out of 2×4s and set them between the posts.

How the walls of a pole barn are framed depends largely on whether the posts are round (A) or square (B). Some 2×4 nailers for plywood or vertical siding can be nailed horizontally to either round or square posts; conventionally framed wall sections can be set between square posts.

A) Round-Pole Wall Framing

2×4

Round pole

Plywood siding

Concrete footing

B) Square-Post Wall Framing

6×6 square post

2×4 stud

Concrete footing

2×4 wall framing

Shed Floors

The building material used for the floor of a shed is often determined by the foundation. For instance, a wood-joist floor frame is usually covered with plywood. However, a timber-frame foundation can have a floor made of brick, gravel, or wood. Also, consider how the out-building will be used before choosing the type of floor. It wouldn't make sense to install carpeting in a tool-storage shed, but it might be a good choice for a kid's playhouse or a writer's studio.

Plywood floor

Plywood is an excellent flooring choice for most stick-built outbuildings. It's easy to install, afford-able, and surprisingly strong. Plus, the large sheets create a very smooth, flat surface, which makes it easy to roll, drag, and push objects across the floor. However, there are dozens of types and thicknesses of plywood. If you install the wrong one, the floor will sag, crack, and eventually rot right out from under you.

Fortunately, picking the right plywood for a shed floor is easy. Just follow this simple rule: Use only ¾-in.-thick, exterior-grade plywood. Anything thinner—even ⅝-in.-thick plywood—won't provide the necessary support.

Standard plywood is manufactured with ordi-nary glue and should never be used for a shed floor. Its thin wood layers (plies) will start to delaminate and blister in less than a year. Use only exterior-grade plywood for the floor. Its plies are glued together with an all-weather, water-resistant adhesive.

I prefer to use ¾-in. ACX plywood for shed floors. The "A" and "C" designations refer to the grades of veneer used on the front and back sur-faces. The "X" identifies it as an exterior-grade product. You can save a few dollars by installing BCX and CDX plywood instead, but they're not as smooth as the A-grade sheets.

The plywood is fastened to the floor joists with 1⅝-in. decking screws or 2-in. (6d) galva-nized ring-shank nails. Most plywood floors

The typical shed floor consists of a 2×6 frame topped with ¾-in.-thick plywood. The 2×6 joists are spaced 16 in. o.c.

Use decking screws or ring-shank nails spaced 8 in. to 10 in. apart to attach the plywood decking to the floor frame.

are left bare, but they'll last longer and clean up easier if you apply two coats of enamel deck paint.

Wood-plank floor

An alternative to a plywood floor is traditional solid wood planking. This type of flooring consists of fat tongue-and-groove planks that measure 1½ in. to 2 in. thick and 6 in. to 8 in. wide.

Although a solid wood floor is considerably more expensive than a plywood one and takes

much longer to install, it does have a couple of distinct advantages. First, a wood-plank floor just looks great, instantly lending old-world charm to an interior space. Thick planks also create a stable, rock-solid floor that won't bounce or bend, even under extreme weight. This type of flooring is a good choice for woodshops that house heavy machines and large workbenches.

The planks are typically installed perpendicular to the floor joists, but you can also lay them diagonally for a more interesting appearance. Just be aware that a diagonal pattern creates 10 to 15 percent more cutoff waste.

Dry-laid brick floor

The main reason I enjoy building a timber-frame foundation is that it affords me the opportunity to put in a brick floor. This style of shed floor is attractive, extremely durable, and fun to install. The installation technique I use is called a dry-laid method because the bricks are simply set down on a bed of sand or crushed stone; they're not adhered with mortar or mastic.

For this type of floor, it's important to use either concrete brick pavers or hardened clay bricks that are specifically designed for use on walkways and patios. Don't use standard clay wall bricks; they're too soft and porous. I'll explain how to install this type of floor in more detail in Chapter 4.

PRO TIP

When using a circular saw to cut plywood, lay the sheet with the best face down. That way, any splintering—called tearout—will occur on the top, or back, surface.

TRADE SECRET

I prefer using tongue-and-groove plywood for shed floors, even though it costs more and is more difficult to install than standard square-edged plywood. The reason? When the plywood sheets are butted together, the tongue on one sheet locks tightly into the groove on the adjoining sheet. The result is a strong, rigid seam that won't sag, even where it spans open spaces between joists.

A timber-frame foundation provides a chance to lay a beautiful brick floor. Set the bricks in a bed of gravel or sand.

PROTIP

Save time and reduce costly measuring errors by cutting a pair of rafters, checking their fit, and then using them as a template to mark the rest of the rafters.

The dry-laid method can be used to install a variety of masonry materials, including cut granite, bluestone, patio blocks, and flagstone. An easier, though less attractive, alternative to laying bricks is to completely fill the frame with concrete, crushed stone, pea gravel, marble chips, or some other loose-fill masonry material.

Slate-on-slab floor

There are few shed floors as strong or as durable as a poured-concrete slab. But let's face it, concrete isn't particularly attractive. One way to dramatically improve the look of a slab floor is to cover it with slate. Again, this probably isn't something you'd do to the floor of a tool-storage building, but it would be terrific for a potting shed or a pool cabana.

Slate is an extremely hard natural stone that comes in various shades of gray, green, blue, and dark red. It's commonly available in square and rectangular tiles and irregularly shaped slabs. The machine-cut tiles are about ¼ in. thick; slate slabs range in thickness from about 1 in. to 1¼ in. Both styles are adhered to a fully cured slab with mortar, but the techniques differ slightly.

Because slate tiles are uniformly sized, they can be installed with thinset mortar. Buy the latex-fortified type and mix it with water. If you can find only standard, unadulterated thinset, then pick up a jug of liquid latex additive and use it, instead of water, to mix up the mortar. The latex will increase the mortar's bond strength and make it much more water-resistant. Slate slabs can't be installed in thinset because their surfaces are too irregular. They must be laid in a thick bed of standard masonry mortar. This method is referred to as a mud job because the mortar is so thick.

Roof Framing

A few rafters and a ridge board are all you need to frame nearly any shed roof—and in some cases, you don't even need the ridge board. Here, we'll take a look at five styles of sloped roofs: gable, shed, saltbox, gambrel, and hip.

Gable and shed-style roofs are the simplest to build, but many folks find saltbox and gambrel roofs more interesting. Hip roofs have a distinctive, refined appearance, but they're rather tricky to build, which may explain why you don't see them on very many outbuildings.

SAFETY FIRST

Avoid taking a tumble with these ladder-safety tips:
- Always set ladders on dry, level ground, never on top of mud, ice, or snow.
- When leaning a ladder against a structure, the distance from the ladder base to the structure should equal one-fourth the ladder's height.
- Don't overreach when standing on the rungs. Keep your hips within the vertical rails of the ladder.

Roof Slope vs. Roof Pitch

TECHNICALLY SPEAKING, the angle of a roof is known as the roof slope. It's calculated by the number of inches it rises vertically for every foot of horizontal run. For instance, a roof with a 10-in-12 slope rises 10 in. for every 12 in. of horizontal run. Most sheds have a roof slope ranging between 6-in-12 and 12-in-12.

The term *roof slope* is sometimes confused with *roof pitch,* which is an older, less-common phrase. Roof pitch is based on the roof's vertical rise divided by its span, or width; it's always expressed as a fraction. For example, if a roof measures 8 ft. in height from the eave to the ridge, then it has an 8-ft. rise. If a shed is 15 ft. wide with a 6-in. overhanging eave on each side, it has an overall span of 16 ft. In that case, the roof pitch is 8 over 16, or ½.

The traditional way to frame a roof is to cut and install the rafters, ridge board, and ceiling joists one board at a time. The advantage of this piecemeal approach is that a single person working alone can easily lift and nail the boards in place. The drawback is that you spend a lot of time clambering up and down ladders, which is tiring and potentially dangerous.

A second roof-framing option is to assemble the roof trusses on the ground, then raise them into place. This technique—used for the sheds in this book—is easier and safer than working from a ladder, but you'll need two or three people to lift the trusses onto the walls.

Gable roof

A gable roof has two sloping planes of equal length. Its familiar A-shaped profile is formed by pairs of common rafters that run at an angle from the tops of the walls up to the peak. A ridge board, if used, runs horizontally between the pairs of rafters where they meet at the peak.

The size and spacing of the lumber used to frame a roof varies, depending on the size of the building. The bigger the building, the larger the boards must be. Most storage sheds are framed with 2×4 or 2×6 rafters and joists, which are

Speed the job of framing the roof by prefabricating the roof trusses on the ground. Then lift each truss into place and screw it to the walls.

ACCORDING TO CODE

If you're building in an area that receives significant snowfall, the shed's roof must satisfy a specific snow-load rating. This building code ensures that the roof is capable of supporting a certain number of pounds of snow per sq. ft. To find out whether your shed roof meets the requirements in your area, contact the local building inspector.

When installing roof rafters or prebuilt trusses, it's important to place each one directly over a wall stud. That way, the weight of the roof will be transferred directly to the foundation. If you set the rafters or trusses between the studs, the weight can bow the plates or eventually crush the walls.

PRO TIP

Although a roof can be framed conventionally with individually installed rafters, a faster and safer way to frame a roof is with simple site-built trusses.

Common rafters

Ridge board

Top wall plate

Ceiling joists

A gable roof's familiar A-shape profile is formed by pairs of common rafters that run at an angle from the tops of the walls to the peak.

TRADE SECRET

Building plans often refer to on center (o.c.) dimensions; for example, rafters are spaced 24 in. o.c. That means the distance from the center of one board to the center of the next measures 24 in., not that there's a 24-in. space between them.

The gable roof of this shed was framed with site-built roof trusses; note that there's no ridge board running along the peak.

spaced 16 in. or 24 in. on center. The ridge board is usually cut from a continuous length of 1×6 or 1×8.

The gable roof on a post-and-beam building is framed with larger timbers, such as 4×4s and 4×6s, but fewer rafters are needed because the hefty framing can support more weight.

When framing a roof with site-built trusses, you don't need a ridge board. Just make sure the rafters are properly positioned before nailing on the plywood sheathing. (For detailed information about building a gable roof with trusses, see the Colonial-Style Shed on p. 154.)

Shed roof

A shed roof is basically half of a gable roof. In other words, it's a single sloping roof plane. You typically see this type of roof on small, narrow structures. Shed roofs can be used on freestanding outbuildings, but they're most commonly found on sheds attached to another structure, such as a house or garage. In that case, a horizontal board called a ledger is fastened to the wall to support the upper ends of the rafters.

Shed roofs are generally built at a shallower pitch than that of gable and saltbox roofs, with a roof slope ranging between 4-in-12 and 8-in-12. When deciding on the angle of a shed roof, take into account where the doors are located. If they're hung on the shed's sidewall, directly under the low edge of the roof, then consider a shallower roof slope. A steeply pitched roof will come down lower, reducing the headroom in the doorway. The only ways to prevent a steep roof from cutting into the doorway are to build taller walls or to make the entire shed narrower—not very practical solutions. (To learn more about building a shed roof, see the Lean-to Shed Locker on p. 76).

Saltbox roof

A saltbox roof is similar to a standard gable roof; both are framed with pairs of rafters that meet at the peak. The main difference is that a saltbox has one roof plane that's slightly longer than the other. This single design change shifts the roof peak off center so it's closer to the front wall, thus creating the distinctive saltbox roof.

Unfortunately, traditional saltbox design has been altered, distorted, and modernized so often that it's sometimes hard to recognize. Today's builders have tried to "improve" the design by changing the angle of the roof slope, raising the walls, and shortening the long roof plane. The result is an ugly, out-of-proportion building. To ensure that your saltbox roof remains true to its colonial roots, follow these two essential design rules. First, frame the roof to a 12-in-12 slope (that's 45 degrees). Second, create the proper proportions by situating the roof peak one-third of the way back from the front wall. (To learn more about building a saltbox roof, see the Saltbox Potting Shed on p. 124.)

Gambrel roof

The classic gambrel roof is easily recognized by its distinctive double-sloping profile. This traditional barn-style design features two short, shallow roof planes that come off the ridge and then break sharply down to longer, steeper slopes.

Gambrel roofs are a bit more difficult to frame than gable, shed, or saltbox roofs, mainly because

SHED-ROOF FRAMING

Basically half of a gable roof, a shed roof is simply a single sloping roof plane. It is found on both attached and freestanding buildings.

Rafter

Front wall

Rear wall

SALTBOX-ROOF FRAMING

Ridge board

12-in-12 slope

Common rafter

Collar tie

1/3

2/3

Front wall

Rear wall

Like a gable roof, a saltbox roof is framed with pairs of rafters that meet at the peak, though one roof plane is slightly longer than the other.

Though more difficult to frame, gambrel roofs are very spacious inside, offer plenty of headroom, and have loft storage above the ceiling joists.

Purlin

Ridge board

Purlin

Common rafters

Top wall plate

When framing a large gambrel roof, it's easier to build trusses on the ground, then lift them into place.

Cutting Corners

IT'S POSSIBLE TO BUILD an entire gambrel roof without cutting a single miter joint. Special metal framing brackets are formed at the precise angles needed to create the double-sloping roof. All you need to do is insert square-cut 2×4s into the brackets and secure them with screws; there are no angles to figure out or gusset plates to install. This shortcut will allow you to frame a gambrel roof in half the time it may ordinarily take. The brackets are available through various mail-order catalogs and at specialty hardware stores. The Fast Framer set shown here is available from Lee Valley Tools®.

These metal framing brackets allow you to build an entire gambrel roof without cutting a single miter joint.

they contain many more parts. However, gambrel roofs are very spacious inside, offer plenty of headroom, and provide loft storage above the ceiling joists.

If you're thinking about building a gambrel roof, remember that the doors must be hung on the end walls because the shorter sidewalls aren't tall enough.

Like most other roof styles, a gambrel roof can be framed piecemeal, one board at a time, or you can assemble trusses on the ground and then lift them into place. (For step-by-step instructions on framing a gambrel roof, see the Gambrel Storage Barn on p. 186.)

Hip roof

A hip roof is essentially a gable roof with four sloping planes. Two planes slant down from the ridge to the sidewalls (as with a standard gable) and one plane at each end slopes down to the top of each end wall. This quadruple-slope design creates an overhanging eave that wraps around the entire building, a feature that's unique to hip roofs.

HIP-ROOF FRAMING

While hip roofs are distinctive looking, they're difficult to frame. You don't see them very often on outbuildings.

Although hip roofs are distinctive looking, you don't see them very often on outbuildings. First, they're more difficult and time-consuming to frame than other roof types. That's because a hip roof has three kinds of rafters: common, hip, and jack. To cut precise miter and bevel joints in these boards is challenging for even an experienced carpenter. However, I used a simplified technique to frame a hip roof for the Garden Shed shown on p. 106.

Stair and Ramp Construction

If the floor of your shed is more than 8 in. or so off the ground, you'll need to build some sort of step or ramp. Keep in mind that you'll often be carrying or wheeling an item in or out of the shed. If the step down is too great, you could easily trip and injure yourself. A properly designed and sturdily constructed set of stairs or

a ramp will make entering and exiting your shed safer and easier.

When designing a set of steps, make the treads at least 10 in. deep and between 7 in. and 8 in. high. It's also very important to make sure all the steps are exactly the same height. Even the slightest discrepancy will create a tripping hazard.

Wooden steps

A set of wooden steps is the best choice when you need to step up two or three times to reach the shed's floor. One option is to build a short set of stairs with a pair of stringers and two or three treads in the traditional manner. Stringers can be cut from 2×12 stock, with two 2×6s used to make each tread. Stairs are usually built a few inches wider than the doorway opening, but in no case should they be less than 36 in. wide.

There are two basic ways to assemble stairs. Stringers can be notched to accept the treads, or the treads can rest on cleats nailed to the inside of the stringers (see the drawing on p. 48). An alternative is to build a couple of simple wooden platforms and stack them on top of each other.

Platform steps are easier to build and more attractive than a typical stringer-and-tread staircase. Set the platforms on solid-concrete blocks to keep them level and stable.

Hip roofs are unique and attractive, but keep in mind that they typically take twice as long to build as a similar size gable roof, and they require about twice as many roof shingles.

Construct wooden steps, platform decks, and ramps from pressure-treated lumber rated for ground-contact use.

PRO**TIP**

To keep a set of stairs or a ramp from sinking into the ground, set it on top of solid-concrete blocks or a few inches of gravel.

TWO WAYS TO BUILD STEPS

Notched Stringer

Tread

Stringer

Cleated Stringer

2×4 ledger

Stringer

Cleat screwed to the stringer

Tread

A two-tier wooden platform creates a warm and welcoming entrance to this workshop. The platform is part staircase, part sun deck.

Make the rectangular base of each platform from pressure-treated 2×6s, then cover it with 2×4s or 2×6s spaced about ¼ in. apart. Build the bottom platform at least 10 in. wider than the top one to create the first step.

Platform deck

A platform deck is similar to platform steps, but much bigger. The idea is to build two or three large wooden platforms of progressively smaller sizes and stack them on top of one another at the entrance to the shed. Again, build the frames out of pressure-treated 2×6s, with joists every 16 in., then fasten down 2×4 or 2×6 decking. For a more visually interesting effect, run the deck boards diagonally in alternating directions from one platform to the next.

A Moveable Ramp

ONE PROBLEM WITH A WOODEN ramp is that it doesn't create the most stylish entrance for an outbuilding. Most homeowners much prefer the look of steps or platforms, but they still occasionally need a ramp to wheel machinery into the shed. The solution is to make a removable ramp from a couple of 2×12s.

To create a smooth transition from the ramp to the shed floor, attach metal ramp fixtures to the 2×12s. The metal fixtures are sold at most home centers, hardware stores, and auto-parts dealers. Bolt one fixture to the upper end of each 2×12 with the hardware provided. Then, whenever you need to wheel in a lawn mower or a wheelbarrow, just set the ramp in the doorway with the lip of the metal fixtures resting on the shed's floor.

The length of a ramp is determined by how high the shed floor is from the ground. The higher the step, the longer the ramp must be. Use the chart at right for guidance when building a ramp for your shed. Keep in mind, however, that longer ramps form gentler inclines that are safer and more comfortable to walk on. To determine how long your ramp should be (L), first determine the height from the doorway to the ground (H).

Make a pair of removable ramps with a set of extruded-aluminum ramp fixtures. Simply bolt each fixture onto the end of a 2×12.

Ramp Height	Ramp Length
12 in.	4 ft. to 4½ ft.
14 in.	4½ ft. to 5 ft.
16 in.	5 ft. to 5½ ft.
18 in.	6 ft. to 6½ ft.
22 in.	7 ft. to 7½ ft.
24 in.	7½ ft. to 8 ft.

What I like best about this type of step is that you can make the platforms virtually any size and in any configuration you like. For example, you can fan out the platforms, offset them by varying amounts, stagger each platform in opposite directions, extend them across the width of the building, or wrap the bottom platform around the corner of the shed. When you're done, you'll have a stylish multilevel entrance deck, not simply a set of stairs.

A stone ramp is suitable for outbuildings that house large tractors, trucks, or farm machinery. However, it might be too bumpy for lawnmowers and bicycles. This ramp is made of rough-cut Belgian block—a.k.a. cobblestone—set on a crushed-stone base.

Arrange several large, relatively flat rocks at the doorway of a shed to create a natural-looking stepping-stone entrance.

Ramps

If you've ever tried tugging a wheelbarrow up a set of stairs, you know why ramps are so valuable. A wooden ramp provides the only safe and sensible way to store lawn mowers, garden tillers, drop spreaders, garden tractors, and many other wheeled devices.

The length of the ramp will depend on the height of the shed floor; the higher it is, the longer you need to make the ramp. As a general rule, a 6-ft. ramp works well in most situations.

Stone steps

I think that natural stone is the best material for building shed steps. It's very attractive, naturally slip resistant, and extremely durable.

You can make a step from a single slab of stone or use several large rocks to create a

2×6 deck boards

Cross blocking

Ledger

Stringer

2 in.
min.

Ramps usually have two or three PT 2×6 or 2×8 support stringers trimmed so they sit flush on the ground and fit against the shed. A 2× ledger supports the upper end of the frame, while cross blocking between the stringers helps support the decking. The ledger is fastened to the shed framing about 2 in. below the door's threshold; the decking is fastened to the stringers with 3-in. galvanized screws (which makes it easier to remove the decking if you need to unbolt the ramp from the shed some day).

TRADE SECRET

Wooden ramps and steps can become slippery, especially when they're wet. To reduce the likelihood of slipping and taking a nasty fall, apply strips of nonslip abrasive tape to the treads. The adhesive-backed tape comes in various widths and is sold by the linear foot at most home centers and hardware stores.

stepping-stone entrance. These methods are especially well suited to situations where you need to step up only once to enter the shed.

Large stones are extremely heavy and virtually impossible for the average person to transport and set in place by himself or herself. The best approach is to buy the stone from a local quarry or masonry supplier and have the dealer deliver and install it.

DESIGN
OPTIONS
ADD-ON ACCESORIES

▲ An attractive—and easy—way to top off a hip roof is with a copper cap and finial. This particular cap is fitted with a green-glass sphere.

▲ A roof extension added to the side of a shed provides shelter for ladders, trash cans, firewood, and other items that don't easily fit into the shed.

▶ An outdoor shower may not be a necessity, but it sure is fun. This one is built onto the side of a pool cabana, providing a convenient place to rinse off before going inside.

▼ A cupola, topped with a weathervane, adds a bit of classical barn architecture. Be aware that the size of the cupola must be in scale with the size of the outbuilding. Too small, and the cupola will look insignificant; too large, and it'll visually overpower the building.

▲ This rustic lean-to ladder trellis is constructed from tree branches. It consists of two 5-ft.-tall vertical poles, and a dozen horizontal rungs. The rungs range in length from about 14 in. to 30 in. They're attached to the poles with galvanized decking screws.

◂ A white-painted trellis creates an eye-catching accent and provides a place to grow climbing plants. This trellis consists of an arch-top lattice panel set between two 4×4 posts.

BUILDING MATERIALS

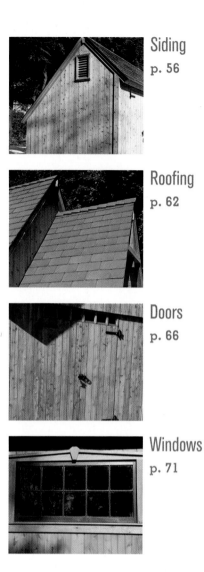

In between the time you design your shed and the time you build it, you'll need to specify all the building materials. The importance of this step is often overlooked, but each decision has an effect on the shed's style, longevity, and cost.

The previous chapter included information about supplies, such as the lumber, foundation materials, and wiring, that don't necessarily influence the look of a shed. This chapter explores the more-visible elements: siding, roofing, doors, and windows.

The building materials you choose will often be based on personal preference, but there are practical considerations as well. You may love the look of a cedar-shake roof, but if you live in a very humid climate or if the site is heavily shaded, wood may not be the best roofing choice; it tends to mold if it remains damp. Wood windows lend a traditional appearance to the building, but aluminum or vinyl-clad windows don't require painting. Consider your choices carefully and you'll be rewarded with a long-lasting shed that's well within your budget. ▶ ▶ ▶

Siding

The type of siding you choose will help define the shed's style as rustic or refined, casual or classic. Siding can be installed horizontally, vertically, or diagonally. Before you make a final decision, however, consider how you plan to finish the siding. Some types take paint really well; others should be stained or coated with clear preservative.

Cost and speed of installation are also valid concerns when choosing siding material. For example, cedar shingles are attractive but rather time-consuming to install. Plywood siding, on the other hand, is reasonably priced and goes up very quickly. Also, some siding types must be nailed to plywood wall sheathing. For other types, you can eliminate the plywood and nail the siding directly to the wall studs.

Wood siding, in all its variations, is by far the most common siding material for sheds and outbuildings. It's relatively affordable, long-lasting, and easy to install. Following are brief descriptions of the six most popular types of wood siding.

Bevel siding

Commonly called clapboards, bevel siding comes in long, thin planks that are installed horizontally on exterior walls. It's called bevel siding because the individual boards are cut with the sawblade tilted at a slight bevel angle to produce planks that are thinner at one edge than the other. As each course of siding is nailed up, the thin upper edge is overlapped by the thicker bottom (butt) edge of the course above it. This is the kind of siding installed on the Saltbox Potting Shed (see p. 124).

Most bevel siding is milled from western red cedar or redwood, two softwood species that

(see p. 124).

Cedar bevel siding (commonly called clapboards) tapers in thickness from 7/16 in. at the butt edge to about 1/8 in. along the upper edge.

There are many varieties of cedar bevel siding, including a rustic, roughsawn knotty grade that's ideal for finishing with stain.

are naturally resistant to rot and wood-boring insects. Spruce, cypress, and pine versions are also available in some regions. Bevel siding comes in several grades, from clear to knotty, and is typically smooth on one side and roughsawn on the other. If you plan to paint the siding, install the smooth side facing out. For a stain finish, install the rough side out; the stain will soak much deeper into the roughsawn surface, and therefore last longer.

PRO TIP

Planning to paint your shed? Save yourself some time and trouble by installing preprimed siding. It comes ready to paint with a factory-applied coat of primer.

TRADE SECRET

To ensure that a paint or stain finish won't peel or fade prematurely, coat the back of each piece of siding before you install it. This technique, known as back-priming, seals the boards and prevents moisture from passing through the siding from the back surface, which can blister the top coat finish. Also, be sure to apply paint or stain on all the exposed edges and the ends of each piece of siding.

Use galvanized siding nails or, better yet, stainless steel nails to fasten bevel siding. Be sure to drive the nails through the plywood wall sheathing and into the wall studs. Otherwise, the nail tips will protrude into the shed's interior and create dozens of pointy, painful hazards.

Tongue-and-groove boards

Pattern siding is a broad family of wood boards that have been machined with various interlocking or overlapping joints. The most popular is V-jointed tongue-and-groove siding. This is the type of siding used on the Colonial-Style Shed (see p. 154).

Each ¾-in.-thick board is milled with a tongue along one edge and a groove along the other. When the boards are nailed up, the tongue of one board fits tightly into the groove of the adjacent board. Generally, an edge on at least one side of the board is also chamfered at a 45-degree angle, creating a decorative V-shaped joint along the seams when the boards are fitted together. This type of siding is typically installed vertically, but it can also be run diagonally or horizontally.

Like other types of pattern siding, tongue-and-groove boards are nailed directly to the wall framing; there's no need to first sheathe the walls with plywood. However, when installing it vertically, you must add horizontal blocking to the wall framing to provide solid nailing support for each board. Pattern siding can simply be face-nailed to the framing, but tongue-and-groove boards look best when the nails are driven through the tongue and then covered by the grooved edge of the next board, a technique known as blind-nailing (see the drawing below).

Tongue-and-groove, V-jointed siding can run horizontally; just be sure to install the boards with the tongue edges facing up.

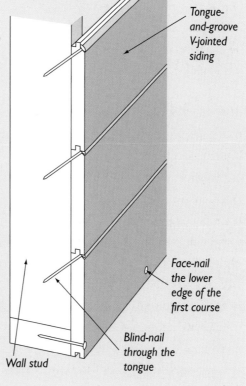

BLIND-NAILING TONGUE-AND-GROOVE SIDING

Tongue-and-groove siding looks best when it's blind-nailed to the wall framing. Drive 2½-in. ring-shank siding nails at an angle through the tongue. The nail heads will be covered by the grooved edge of the next siding course.

Tongue-and-groove V-jointed siding

Face-nail the lower edge of the first course

Blind-nail through the tongue

Wall stud

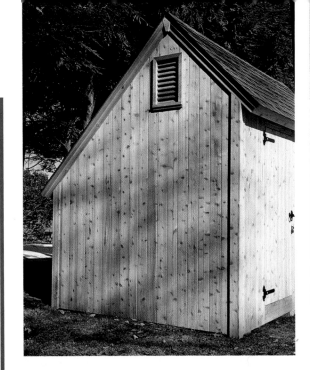

This shed is sided with tongue-and-groove, V-jointed cedar 1×6s. The knotty boards are finished with clear wood preservative.

Channel siding can be installed vertically, diagonally, or horizontally. The roughsawn boards are available in widths of up to 10 in.

CHANNEL-SIDING INSTALLATION

Channel siding

Wall stud

1½-in.-wide rabbet

1½ in.

Fasten channel siding to the wall framing by face-nailing each course 1½ in. up from the bottom edge with 2-in. ring-shank siding nails (use one nail for siding up to 6 in. wide; two nails per stud for siding 8 in. and wider). The narrow rabbet on the bottom edge partially overlaps the larger rabbet on the top of the adjacent board, effectively holding the board in place.

Face-nail along the lower edge

1-in.-wide channel

½-in.-wide rabbet

Most tongue-and-groove siding is milled from western red cedar or redwood and has either a smooth or a roughsawn surface. It's readily available in 4-in.-, 6-in.-, and 8-in.-wide planks, with the 6-in. size being the most common. You can also buy tongue-and-groove boards made from untreated pine, but be aware that this product is very susceptible to rot—especially on the boards closest to the ground. If you choose untreated pine siding, protect it with stain or paint, then reapply the finish at the first sign of water damage.

Channel siding

This type of pattern siding is popular because of its versatility and rustic appearance. It's most often cut from knotty, roughsawn cedar and available in 6-in., 8-in., and 10-in. widths. Choose either 6-in. or 8-in. siding for small to medium-size sheds; select 10-in. boards for large outbuildings. Channel siding can be nailed up vertically or diagonally, but it is often installed horizontally to mimic traditional clapboards.

The ¾-in.-thick boards have a modified rabbet joint cut into each edge. (A rabbet is simply an L-shaped notch running along the edge or end of a board.) A ½-in.-wide rabbet is milled from the rear surface along one edge, and a 1½-in.-wide rabbet is milled from the front surface on the opposite edge. When the boards are installed, the narrower rabbet on one board partially overlaps the larger rabbet on the adjacent board, resulting in a 1-in.-wide reveal, or channel, between the planks (see the drawing at left).

Board-and-batten siding

One of the oldest types of wood siding, board-and-batten siding consists of wide boards and narrow wood strips, called battens. The boards are nailed vertically to the wall framing, then the battens are installed to conceal the seams between the boards. Note that this type of siding can only be installed vertically.

Unlike other types of wood siding, there's no standard size for board-and-batten siding: You can make it from virtually any size lumber. However, the battens are usually cut from 1×3s and the vertical boards are cut from stock ranging from 1×8s to 1×12s. Again, narrower boards look best on smaller sheds; wider boards are used for larger buildings. For best results, cut the siding from a rot- and insect-resistant wood, such as cedar, redwood, or pressure-treated lumber. When you install the boards, leave about a 1-in. space between them so they can expand without buckling. If the boards are 6 in. or narrower, secure each one with a single row of nails driven through the middle of the board. If they're 8 in. or wider, attach them with two rows of nails.

After the boards are in place, fasten the battens over the gaps between the boards with a single row of nails. Make sure the battens overlap the boards by a minimum of ½ in. to provide adequate coverage in case the boards shrink (see the drawing at right).

Cedar shingles

A good way to capture the look of a quaint country cottage is by siding your outbuilding with cedar shingles. The individual shingles

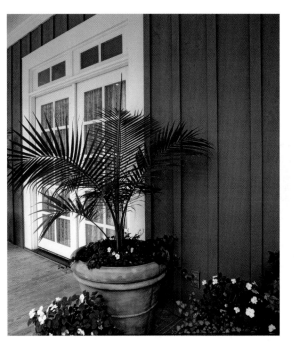

Capture the casual cottage look with board-and-batten siding, which consists of wide vertical boards and narrow wood battens.

To prevent siding from splitting as you nail it, first bore a pilot hole with a drill bit that's slightly smaller in diameter than the nail shaft.

BOARD-AND-BATTEN SIDING

1-in. minimum space

½-in. minimum overlap

1×3 batten

1×6 board

Leave a 1-in. space between the boards and fasten the battens over the gaps with a single row of nails. Make sure the battens overlap the boards by a minimum of ½ in. on each side. Board-and-batten siding is particularly well suited for roughsawn "green" lumber. The single-row nailing pattern allows the boards to dry out and shrink without splitting. Green lumber can be purchased directly from a sawmill for less than the kiln-dried lumber found at lumberyards and home centers.

Nail through the center of the batten and into the horizontal wall framing.

These red-cedar shingles are finished with clear wood preservative to help retain the wood's natural beauty and auburn color.

CEDAR-SHINGLE INSTALLATION

Wall stud

Plywood
wall
sheathing

16-in.
cedar
shingles

Felt
underlayment

6-in.
exposure

Starter course

The first course is ½ in.
lower than the starter
course to form a drip edge.

Cedar shingles must be nailed to plywood wall sheathing. The nails in each row are concealed by the next course. A starter course of shingles is installed along the bottom of the wall to flare out the first row of shingles.

Cedar shingles (bottom) are thinner and smoother than hand-split shakes (top), which are typically thicker and more roughly cut.

Shingle Styles

ONE REASON YOU MAY want to consider using shingles is that they're available in a wide variety of custom-cut shapes, allowing you to add dramatic designs to an otherwise ordinary façade. Available from the same mills that sell standard shingles, these Fancy-Butt shingles comes in nine shapes, including diamond, round, half cove, hexagonal, arrow point, and fish scale.

are installed in overlapping courses, which shed rainwater like water off a duck's back. The nails in each course are concealed by the shingles in the course above it, resulting in a very neat, natural appearance.

For most outbuildings, shingles aren't as popular as other types of wood siding. For one thing, cedar shingles are rather time-consuming to install and relatively expensive. Also, they can't be nailed directly to the wall framing; you must first install plywood sheathing. Finally, after the shingles are installed, there will be hundreds of nails poking through the walls and into the spaces between the wall studs. To keep from getting pricked accidentally, you'll have to cover the walls on the inside of the shed with plywood, wallboard, pegboard, or some other rigid material.

Like other types of wood siding, shingles can be finished with paint, clear wood preservative,

Textured plywood siding comes in 4-ft. by 8-ft. sheets and provides the quickest, easiest way to cover large wall sections.

or semitransparent or solid-color stain. It's not important which finish you apply, just that you apply one. Shingles last much longer and curl and split less when they're protected from the elements with a coat of finish.

Looking for a more rustic, less refined appearance? Consider installing hand-split cedar shakes. Shakes are similar to shingles, but they're much thicker and more heavily textured Shakes can also be installed with a greater exposure to the weather, so you'll need fewer shakes than shingles to cover the shed.

Plywood siding

There are two valid reasons for siding your shed with plywood: It's economical and it covers a lot of area very quickly. Plywood siding is a $\frac{5}{8}$-in.-thick exterior-grade product that's commonly available in 4-ft.-wide by 8-ft.-long sheets. It can also be special-ordered in 9-ft.- and 10-ft.-long panels.

Plywood siding comes in several styles, including roughsawn, primed, and unprimed wood, as well as something that resembles reverse board-and-batten siding. However, the most popular style by far is grooved plywood siding (though nearly everyone refers to it as T-1-11®, the trade name of grooved siding manufactured by Georgia Pacific).

Reverse board-and-batten plywood siding has a rough-textured surface machined with broad, widely spaced grooves.

This type of siding has a roughsawn surface that features a series of equally spaced $\frac{1}{4}$-in.-deep by $\frac{3}{8}$-in.-wide grooves. You can buy it with the grooves spaced either 4 in. or 8 in. apart. The style to choose depends on the size of your building. The 4-in. design looks best on smaller structures; larger outbuildings can accommodate the wider 8-in. pattern. I installed 4-in. grooved siding on the small Lean-to Shed Locker (see p. 76). The large Gambrel Storage Barn is sided with 8-in. grooved plywood (see p. 186).

Plywood siding is nailed directly to the wall framing. Shallow rabbets, similar to those on

Grooved plywood siding is commonly available with the ⅜-in.-wide grooves spaced either 4 in. (shown) or 8 in. apart.

channel siding, are milled into the edges of the sheets. When the panels are installed, one sheet overlaps the next and creates a tight, water-resistant joint. To ensure that the plywood panels won't pull free over time, I like to apply a bead of construction adhesive along every stud before installing the plywood siding. That way, if the framing lumber shrinks slightly or if the nails work themselves loose, the siding will remain fast.

Roofing

During the initial design phase of your outbuilding, you probably won't spend a lot of time fretting about which type of roofing to install. That's certainly understandable, as so much forethought must go into selecting the siding, windows, and paint color. However, don't underestimate the impact a roof can have on the style and appearance of an outbuilding.

There are a wide variety of roofing materials available, but most of them aren't very well suited for do-it-yourself installation. Here, I'll discuss the three most appropriate roofing options for the backyard builder, including two reliable standards—asphalt shingles and cedar shakes—and a modern version of an old classic: slate.

Asphalt shingles

To say that asphalt shingles are popular is an understatement of immeasurable proportions. I'd estimate that a vast majority of outbuildings—probably close to 95 percent—are covered with asphalt roofing. And why not? It's affordable, easy to install, readily available in a wide array of colors, and surprisingly durable; many brands carry up to a 40-year warranty.

There are two basic kinds of asphalt shingles: organic-based shingles and fiberglass-based ones. The organic type is usually less expensive and available from more manufacturers; fiberglass-reinforced shingles last longer, are lighter in weight, and are more fire resistant. You can find very good—and not-so-good—versions of both kinds at most lumberyards and home centers.

There are also two main categories of asphalt shingles: three-tab shingles and architectural-style shingles. Standard three-tab shingles represent a very basic, generic style. Each shingle is a single layer thick, with two narrow slots cut into it to create the three tabs.

SAFETY FIRST

Before climbing onto any roof, make sure the surface is dry and free of debris. Never work on a roof that's wet, icy, or covered with snow or dew. Lash the top of the ladder to the building with a rope or chain so it can't slide away. Finally, wear soft-soled shoes to reduce slipping.

Fiber-Cement Siding

FIBER CEMENT IS A UNIQUE SIDING MATE-RIAL THAT looks like wood but is actually a mixture of cement, sand, and cellulose fibers. It has been around for nearly 20 years and is slowly gaining popularity with professional homebuilders. However, it's seldom used on outbuildings, even though, in many ways, it's a near-perfect choice. (I installed fiber-cement siding on the Garden Shed shown on p. 106.)

Fiber-cement siding won't rot, crack, delaminate, or burn. It's highly resistant to moisture, including salt air and salt spray. Wood-boring insects hate it, and it holds paint better than almost any other type of siding does. In most cases, it's even less expensive than quality wood siding.

However, fiber-cement siding isn't the world's most user-friendly building material. The siding is much heavier than wood and more difficult to nail because it's so hard. It's also nearly impossible to cut with ordinary tools. You can use a portable circular saw fitted with an appropriate carbide-tipped blade, but it'll spew out a cloud of dust thick enough to shut down an airport. You can also score and snap it with a carbide-tipped scoring tool (doesn't that sound like fun?), but the best way to cut fiber cement is with a pair of electric or pneumatic shears. The shears are very expensive to buy, but you should be able to rent a pair at a well-stocked tool rental outlet.

Fiber-cement siding comes in $5/16$-in.-thick by 12-ft.-long planks and ranges in width from 5 in. to 12 in. The planks resemble wood clapboards; the material is also available in various shingle-style planks that go up quickly but look like individual shingles. Some manu-facturers also offer styles that mimic grooved plywood siding, stucco, and specialty shingles with rounded, scalloped, and clipped-corner designs.

Fiber-cement siding is embossed with a realistic wood-grain texture that looks very much like roughsawn bevel siding.

What appear to be individual wood shin-gles installed with staggered butt edges are actually 4-ft.-long fiber-cement planks.

These attractive, rounded shingles are just one of the many varieties of specialty shapes now available in fiber cement.

WHAT CAN GO WRONG

Don't wait until the day you nail down the ply-wood roof sheathing to go buy the shingles. Most lumberyards and home centers stock only three or four of the most popu-lar shingle colors. If you want something a little out of the ordinary, such as forest-green shingles, it'll probably have to be special-ordered, which could take anywhere from a couple of days to a few weeks. I usually order the roofing on the same day that I pick up the framing lumber.

PRO TIP

Don't lay asphalt shin-gles on extremely hot, sunny days. Walking on softened shingles will wear away surface gran-ules and shorten the life of the roof.

Roof Language

HERE ARE FIVE COMMON roofing terms that every backyard builder needs to know:

- **Sheathing.**
  The plywood or solid wood layer to which the shingles are nailed.
- **Drip edge.**
  The metal flashing applied to the roof edges to deflect rainwater runoff.
- **Building paper.**
  An asphalt-saturated felt underlayment, sometimes called tar paper, that's stapled to the sheathing.
- **Starter course.**
  A single layer of shingles installed along the edges and ends of the roof prior to the first course.
- **Exposure.**
  The amount of shingle left exposed to the weather, usually about 5 in.

Asphalt roof shingles come in two basic types: three-tab shingles (top) and architectural-style shingles (bottom), which are also called laminated shingles.

The variegated gray and black tones of these architectural-style roof shingles were designed to look somewhat like a traditional slate roof.

Architectural-style shingles, which are also commonly called laminated shingles, consist of two strips of asphalt roofing, one laid on top of the other. This type of shingle has a solid bottom piece and a top strip notched with widely spaced, dovetail-shaped tabs. The two strips are laminated together at the factory. When installed, the laminated construction forms a heavily textured surface with deep shadow lines.

Architectural-style shingles are a dramatic improvement over three-tab shingles in terms of visual impact. The shingles are available in several muted colors that, in many cases, mimic traditional roofing materials. For example, the tan tones resemble weathered cedar shakes and the gray-black shingles look somewhat like slate.

Cedar roofing

Many roofing materials are more affordable, longer lasting, and easier to install than a cedar roof, but none can compare to its natural beauty and distinctive texture. The one disadvantage of any wood roof, of course, is that it offers very little resistance to fire, unless it's treated with a fire retardant.

There are two basic types of cedar roofing: shingles and shakes. Shingles are thinner, smoother, and more uniform. Shakes are thicker and rougher. Some shakes are sawn from logs; others are hand-split for a more rugged, uneven surface.

A shingle roof has a smart, elegant appearance, with clean lines and a low profile. The Saltbox Potting Shed (see p. 124) has a cedar-shingle roof. A shake roof is more rustic and robust-looking. Its surface has a lot of character and deep shadow lines.

A cedar roof is normally installed over spaced sheathing, not plywood. Spaced sheathing—also called open or skip sheathing—is simply a series

ACCORDING TO CODE

For fire safety reasons, some building departments have banned the use of wood roofs. The original building code was written to protect house roofs, but the code is typically applied to all structures, including outbuildings. If you're considering a wood-shingle or shake roof for your shed, be sure to get approval from the local building inspector.

Shingles are uniformly sawn to about 3/8 in. thick, resulting in a stylish, low-profile roof that casts very thin shadow lines.

Hand-split cedar shakes have a rough texture and very thick, uneven butt edges that cast long, deep shadows across the roof.

of 1×4 slats nailed across the rafters. The slats are spaced a few inches apart, which allows air to circulate behind the shingles or shakes to keep them dry. That's the key to a long-lasting wood roof; if the shingles can't breathe, they'll stay damp and quickly deteriorate.

Western-red-cedar shingles and shakes are widely available nationwide. In some areas you may also find eastern-white-cedar shingles, which are fine for sidewalls, but stick with western red cedar for roofing. It simply lasts longer and outperforms any other type of cedar.

Cedar shingles are available in several grades. For sidewalls and roofs, order Number 1 Blue Label® shingles. These premium-grade shingles are perfectly clear (no knots) and cut from all-heartwood for optimum decay resistance. They're available in 16-in., 18-in., and 24-in. lengths; the 16-in. size is suitable for most roofing jobs.

There are four premium grades of cedar shakes. My favorite choice for roofing is a grade designated as Certi-Split® hand-split shakes. It has a rough-split face, but the smooth-sawn back makes installation easier.

Faux slate

Real slate roofs are durable and beautiful, but they aren't typically used for shed roofs because they're difficult to install and expensive. Dura Slate® Roofing System shingles are a new version of this ancient roofing material. This imposter resembles real slate—even up close—but it's made of a polymer compound similar to hard rubber.

The 12-in. by 18-in. shingles won't rot, dent, or split. They are flexible enough to withstand rough handling, are extremely water-resistant, and come with a 50-year warranty. The shingles

Looking very much like real slate, these 12-in. by 18-in. Dura Slate Roofing System shingles are made of tough polymer similar to hard rubber.

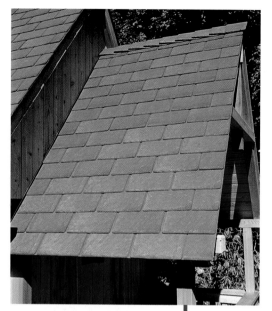

The realistic muted gray colors and embossed surface of this faux-slate roof are almost indistinguishable from real slate.

are resilient but can easily be cut with a utility knife—just score and snap them. Faux-slate shingles are installed in a manner similar to that of asphalt shingles: They are fastened to plywood sheathing with standard roofing nails. Although faux slate isn't cheap—it is usually two to three times more expensive than asphalt shingles—the installation time, tools, and degree of difficulty are basically the same for both types of roofing.

I installed a Dura Slate roof on the Lean-to Shed Locker (see p. 76) and noticed that the shingles were much easier to cut and nail after they were warmed by the sun for a few minutes. When cold, the polymer shingles turn hard and are much more difficult to score with a utility knife.

The one telltale sign that Dura Slate is not the real thing is that the shingles are a little too perfect—a bit too uniform in size, shape, and color. Real slate is slightly more irregular, with scaly surfaces and corners of varying shapes.

Doors

There are no hard and fast rules to follow when it comes to selecting a particular type of door for an outbuilding. As long as it opens smoothly and latches securely, it doesn't much matter what style the door is. Begin your search for the right door by taking a very practical approach. First, decide whether you need a single or a double door. A single door is adequate for most sheds, but if you're going to store a garden tractor, boat, or trailer, you'll need to install a pair of double doors.

Next, choose between a hinged swinging door and a sliding door. Swinging doors are easier to install and the hardware costs much less, but sliding doors glide completely out of the way and they won't accidentally slam shut. Another important consideration, especially if you live in a colder climate, is that hinged doors close more tightly and do a better job of blocking wind than sliding doors do.

Also, think about whether you want the door to discreetly blend in with the shed or to stand out as an architectural accent. To make the door inconspicuous, simply paint or stain it to match the siding. In fact, if you build it from the same material as the siding, it'll virtually disappear. If you'd prefer to create an eye-catching focal point, paint the door a bright color, use ornate wrought-iron hardware, or install elaborate architectural trim around the door opening.

Following are descriptions of four types of doors that you can readily build from commonly available lumber. There are no tricky joints to mill or complicated assembly sequences; each can be built with nothing more than a cordless drill/driver and a portable circular saw or tablesaw.

However, if you don't have the time or inclination to build a door from scratch, don't worry about it. Go out and buy one. Most

Security Measures

SURFACE-MOUNTED hinges are the easiest type of door hinges to install on an outbuilding, but they don't offer much in the way of security. Even with the doors padlocked, someone can get into the shed by simply removing the hinge screws. To thwart such thievery, replace at least one screw in each hinge leaf with a carriage bolt. If necessary, enlarge the hole in the hinge to accept the bolt, then add a washer and hex nut on the inside.

home centers and lumberyards stock several styles of prehung doors made from wood, steel, and fiberglass.

Batten doors

Carpenters have been building batten doors for more than five centuries. This style of door is still popular today—especially with do-it-yourselfers— because it looks great and is very easy to build.

A batten door consists of little more than a few vertical boards joined together with wooden strips, called battens. The battens are usually attached to the back of the door in an X- or a Z-shaped pattern, which reinforces the door and keeps it from sagging. You can make a door from standard, square-edged 1× boards, but

it'll be much stronger and more attractive with tongue-and-groove, V-jointed cedar 1×6s. Plus, should the boards shrink slightly, no gaps will appear between them because of the interlocking tongue-and-groove joints.

It's pretty easy to build a batten door. First, start by cutting all the 1×6s to length. Then lay the boards face down on a flat surface and clamp them together. Next, saw off the tongue and groove from enough of the 1×6s to make the battens. Lay out the battens on the door as shown in the drawing on p. 68, cut them to length, apply a bead of construction adhesive to each one, and screw them to the back of the door. For the average-size door, I recommend attaching three horizontal battens—one each at the top, center, and bottom—and two diagonal battens.

This handsome batten door is made from tongue-and-groove V-jointed boards, the same 1×6 boards used for the shed's siding.

A Grand Opening

THE OWNER OF THIS SHED WANTED a traditional-looking outbuilding, but he also wanted easy access to the interior for storing an antique tractor, lawn mower, and snow blower. The solution was ingeniously simple: Four doors were hung on the rear of the building, creating two 8-ft.-wide bays. This four-door, rear-drive design is also suitable for workshops: Open up the doors to load in materials and to move machinery in and out. During warm weather, the doors can be opened for fresh air and light.

Like most backyard buildings, this cedar colonial has a pair of swinging doors on the front façade (left). But this shed is hiding a secret: There are four doors mounted to the rear wall (right), which create a wide-open, easy-access entrance to the interior.

The back view of this batten door reveals the positions of the three horizontal battens and the two diagonal Z-battens.

1³⁄₈ in.

1³⁄₈-in. space accommodates the astragal strip on the mating door in double-wide openings.

Tongue-and-groove 1×6

2 in.

4¹⁄₄-in.-wide batten (ripped from a 1×6)

2 in.

The rear view of a typical Z-batten door reveals three horizontal battens and two diagonal ones. The door panel is made from several tongue-and-groove boards.

A plywood door is made from a single plywood panel, which is then trimmed with a perimeter face frame cut from 1×4s.

Plywood doors

This style of door is the quickest and easiest type to build. It's simply a piece of plywood cut to fit the door opening, then framed with 1×4s to strengthen and stiffen the door Not surprisingly, plywood doors are usually installed on sheds covered with plywood siding. And, of course, the same plywood used for the siding can be used for the doors.

These doors are often made with ⁵⁄₈-in.-thick grooved plywood, but plain (not grooved) plywood siding with a roughsawn texture works just as well. For the frame, I like to use a weather-resistant wood, such as cedar or redwood.

Dutch doors

A Dutch door is easily recognized by its unique split personality: The top and bottom halves of the door swing independently of each other. The bottom section is often closed for a modicum of privacy and security, whereas the top half is left open for fresh air and neighborly chats. The two halves can also be locked together to operate as a standard swinging door. Although you probably wouldn't hang a Dutch door on a basic storage building, it'd be the perfect choice for adding a little old-world charm to a woodshop, backyard stable, or potting shed.

Dutch doors originated in (surprise!) the Netherlands during the early 1600s. They were first installed in front entryways before being relegated to secondary doorways at the side and rear of the house. Nowadays, Dutch doors are rare, though you may occasionally see one on a horse barn, garage, or roadside produce stand.

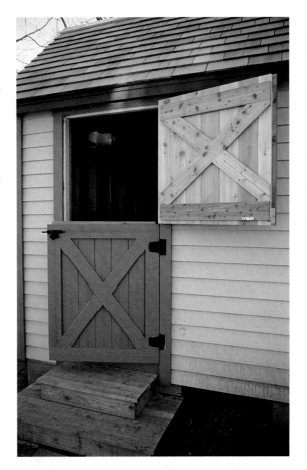

Form and function come together in this charming Dutch door. Note the use of X-battens to create the crossbuck design.

TRADE SECRET

Here are three important rules to follow when making a plywood door:

- Use exterior-grade plywood that's a minimum of $5/8$ in. thick.
- Attach the wood frame with weather-resistant galvanized or stainless steel screws.
- To reduce the likelihood that the plywood will bow or twist, apply two coats of stain or paint to both sides and to all four edges of the door.

PLYWOOD DOOR

4 in.

1×4 stile

$5/8$-in.-thick grooved plywood door panel

Approximately 32 in.

1×4 rail

One of the easiest of all shed doors to build, a plywood door is simply a grooved plywood panel with a 1×4 face frame screwed on from the back. Attach the face frame to the front of the plywood with construction adhesive and decking screws. Drive screws from the back so the screw heads won't be visible on the front. Run the 1×4 frame around the perimeter of the plywood panel, then attach a horizontal 1×4 across the door at or slightly below the center point.

Don't make the door until after the shed has been framed and the siding has been installed. Then custom-build the door to fit the opening precisely.

The scarcity of these double-swingers isn't because people don't like them; in fact, most people love Dutch doors. You don't see them very often because Dutch doors are expensive to buy, few carpenters want to be bothered making them, and even fewer homeowners know how to build one. I can't do anything about the first two reasons, but I can explain a couple of relatively simple ways to make your own Dutch door.

The first method is based on the design of a traditional batten door. Fasten together several tongue-and-groove boards with battens. Then saw the door in two and hang each half with a pair of hinges. For an out-swinging door, consider putting battens on both sides. That way, when the bottom half is closed and the top half is open, the two visible surfaces will

The sliding batten door on this woodshop features a large window that sports a decorative grille made of ⅝-in.-dia. rebar.

match. (For more information on building a Dutch door, see p. 146.)

Here's an even easier way to build a Dutch door. Buy a solid-core, flush-panel exterior door and cut it in half. Protect the freshly cut ends by gluing on thin wood strips, then paint or stain the fronts, backs, and all the edges to seal out moisture. To create a little visual interest, rout shallow V-grooves in the door (spaced about 4 in. apart) or trim it with faux battens made from 3½-in.-wide strips of ¼-in. plywood.

Sliding doors

Long before the advent of sectional roll-up doors, nearly every carriage house, horse barn, and garage was outfitted with sliding wooden doors. Today, sliding doors still make sense for many larger outbuildings. They glide open with very little effort, can't be blown shut by the wind, and cover larger openings than do standard swinging doors. Because sliding doors fit over—not into—the doorway, you don't have to spend any time trimming and fussing to make the door fit precisely.

You may want to consider a sliding door if you live in an area that receives significant snowfall. Unlike an out-swinging door, a sliding door can be opened the morning after a blizzard without your having to shovel the snow out of the way first. That's a big advantage, especially if you store your snow blower in the outbuilding.

A sliding door is basically just a slab of wood suspended from an overhead steel track. Virtually any style of door can be used with a sliding track, including plywood, solid-core, and batten ones. The door is attached to the track with a roller mechanism. Two roller assemblies (called trucks) are required for each door. Quite a few companies make sliding-door hardware, which come in only two basic types: trolley hanger and

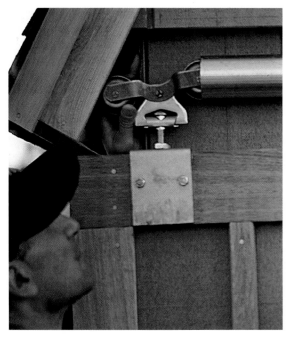

Sliding doors are hung from rollers that glide on a track. The round rollers of this cannonball hanger fit into a tubular track.

cannonball hanger. The trolley hanger has a wheel assembly (similar to a roller skate) that fits into a square or rectangular track. The cannonball hanger has a ball-shaped roller mechanism that fits into a round tubular steel track. Both types work well, but I've found that the cannonball hanger rolls a bit more smoothly and won't jump off the track.

Whichever type of sliding-door hardware you choose, make sure it can be easily adjusted after the door is hung. Check the installation instructions or call the manufacturer. It's important that the hardware have both vertical and lateral adjustments.

Windows

Natural light, fresh air, and views are just three reasons for installing windows in sheds. They also add character and style to an otherwise ordinary façade. There are dozens of window types and

sizes from which to choose. Most are made of wood, aluminum, or vinyl. Clad windows are built from wood but have an exterior covering— or cladding—of aluminum or vinyl.

The type of outbuilding you construct will dictate the size, type, and number of windows you should install. For example, a single double-hung wooden window may suffice for a small storage shed, but a woodshop or artist's studio will be better served with several tall, light-grabbing casements.

Generally speaking, any window that you'd put in a house can be installed in an outbuilding. Here, I'll concentrate on the three types most commonly used for backyard buildings: wooden barn sashes, aluminum sliders, and fixed transoms.

Window Words

HERE ARE BRIEF DESCRIPTIONS of six common types of windows:

• Double hung.
A window with two sashes, each of which slides up and down.

• Single hung.
A two-sash unit, but only the bottom sash is operable.

• Casement.
A crank-operated window with a side-hinged, swing-out sash.

• Hopper.
A small, usually horizontal window that hinges at the bottom and tilts in at the top.

• Awning.
A top-hinged window that swings out at the bottom.

• Picture.
A large, flat, fixed (non-opening) window, often flanked by narrower casement or double-hung units.

PROTIP

Before nailing a window in place, close and lock the sash to help hold the window frame square. Also, make sure the frame is level and plumb.

A barn sash is a true divided-lite window, with single glazing set in an unfinished pine frame. Shown at right is a 2-ft. by 2-ft. unit.

To add a little style and visual interest, this 2-ft. by 4-ft. barn sash window was painted dark green to contrast with the stained cedar siding.

Wooden barn sashes

As its name implies, a wooden barn sash is used in barns, stables, chicken coops, and other farm buildings. It's the most rudimentary and affordable of all window types. Technically, a barn sash isn't even a window; it's just a window sash comprised of glass panes set in a simple pine frame.

Before installing a barn sash, you'll need to build a window frame for it. I typically build frames out of 1× cedar, redwood, or pressure-treated lumber and add stops (narrow wood strips) to the inside of the frame to hold the sash in place. For ventilation, I like to allow the sash to tip in toward the shed interior.

Barn sashes are commonly available at lumberyards and farm supply stores in 2-ft. by 2-ft. and 2-ft. by 3-ft. sizes. It's a good idea to apply a coat of paint or stain to all wooden surfaces of the sash, then let it dry thoroughly before setting it in the frame. Note that barn sashes look best when they're painted a color that contrasts with the siding.

Aluminum sliders

Many architectural purists will scoff at the thought of putting a modern aluminum window on a traditional-style outbuilding. And they're right. These windows may look out of place on a colonial-style saltbox sided with cedar shingles. However, for many other outbuildings, aluminum sliders are ideal.

I like this style of window because the aluminum frame never needs painting, the product comes with insect screens, and, when it's open, the sash is contained within the frame—it doesn't swing out or lean into the shed. Aluminum sliders come in dozens of sizes, but the two most popular models for outbuildings are 2-ft. by 4-ft. and 2-ft. by 6-ft. units. Sliders are also easy to install: Just set the window in the rough opening from the outside, then have a helper on the inside drive a couple of screws through the aluminum jamb and into the wall framing on each side of the window.

Sliding aluminum windows are very durable, never need painting, and come in several sizes. Shown at left is a 2-ft. by 6-ft. unit.

Transom windows add some light to a shed's interior but are usually installed as an architectural accent for a doorway.

Fixed transoms

This long, narrow window is installed over a doorway. It doesn't open for ventilation and emits a relatively small amount of daylight, but it's a valuable asset just the same.

A transom window adds a touch of elegance and visual interest to any entryway. It's one of the many small details that you can incorporate into your shed design to make it unique. Transoms typically range from about 8 in. to 12 in. high and are as wide as the doorway. They look best when installed over a pair of double doors, but there's no reason you can't put one over a single door.

Transom windows are available from most major window manufacturers, but it's much cheaper to simply make one from scratch. Start by cutting a wooden frame from 1× stock; rip the pieces to about 2 in. wide. Rout a $3/16$-in.-deep by $7/8$-in.-wide rabbet in the back of the frame pieces. Assemble the frame, then order a piece of double-strength glass to fit in the rabbets. Secure the pane to the frame with clear silicone adhesive. You can install the window as is or cut short vertical wood strips to create the look of a divided-glass sash.

DESIGN
OPTIONS
DOOR DESIGNS

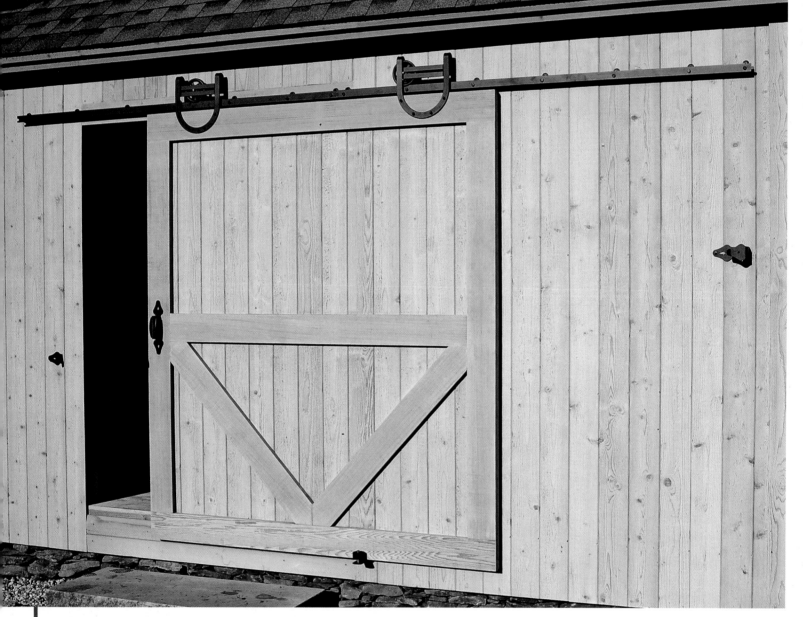

▲ **Sliding doors are popular because they glide smoothly and allow total access to the doorway opening. And sliding doors can't be blown shut by the wind.**

▲ A pair of wide swinging doors, salvaged from an old dairy barn, grace the front of this colonial-style storage shed.

◄ The unique doors on this shed were made from an old pair of shutters. Each 16-in.-wide shutter is attached with three strap hinges. The best place to find old shutters is at an architectural salvage yard.

▶ This attractive batten door is built of tongue-and-groove cedar 1×6s. The casing around the door is painted a contrasting color to make the door stand out from the siding.

▲ If you don't want to build a door, then purchase a prehung exterior door. This shed has an insulated steel door; fiberglass and wood models are also available.

LEAN-TO SHED LOCKER

One of the primary benefits of building your own shed is that you can design it to satisfy a specific storage need. This compact shed is only about 2 ft. deep by 6 ft. wide, but its shallow, reach-in design is ideal for storing lawn and garden tools.

This style is known as a lean-to because it's built against an existing structure, such as a house, garage, or larger outbuilding. It features grooved plywood siding, double-wide doors, and a beautiful faux-slate roof. The timber-frame foundation is inlaid with red-brick pavers.

Note that the timber frame extends 2 ft. beyond the shed to create a small brick "patio." This clever detail ensures that you'll always have a dry, level place to stand while retrieving a tool. The interior is outfitted with six wooden shelves, a door-mounted storage rack, and a large perforated hardboard panel for hanging long-handled tools. (To order a set of building plans for the Lean-to Shed Locker, see Resources on p. 214.) ▶ ▶ ▶

Timber-Frame
Foundation
p. 78

Wall
Framing
p. 85

Roof
Framing
p. 89

Siding
and Trim
p. 91

Roofing
p. 94

Doors
p. 98

Interior
Shelves
p. 101

Timber-Frame Foundation

At the base of this 2-ft. by 6-ft. shed locker is a 4-ft.-wide by 6-ft.-long timber-frame foundation made of pressure-treated 4×6s. The four timbers are joined together with overlapping half-lap joints, then set on a bed of crushed granite. The stone base is necessary to prevent the timber frame from sinking into the soil. Once the frame is installed, 2⅜-in.-thick concrete-brick pavers are laid to create a hard-wearing, all-weather floor.

Assemble the frame

I used pressure-treated 4×6s rated for ground contact to build this timber-frame foundation, but 4×4s or 6×6s would have worked just as well. Because the ground was relatively flat, I needed only one course of timbers. However, if your building site slopes more than 5 in. or so, you'll have to stack two or more timbers at the lowest end. If the ground slopes only 3 in. or 4 in. across the width of the foundation, you should be able to level the frame by partially burying the high end in the soil.

The first step is to cut the timbers to length to create the proper size frame. However, the exact frame dimensions are based on the size of the brick pavers used for the shed floor. That's because you want to avoid having to cut any bricks. It's much easier to adjust the size of the frame to accept an exact number of whole bricks. The frame doesn't need to be precisely 4 ft. wide by 6 ft. long; it can be a few inches larger or smaller.

For example, the bricks I used to create the floor for this shed locker measure 3⅞ in. wide by 7⅞ in. long. To accept a whole number of bricks, I cut the frame to 46¾ in. wide by 78⅜ in. long. If you're using a different size brick, you'll have to calculate the dimensions of the frame before cutting the 4×6 timbers.

Sizing a Timber-Frame Foundation

ONE OF THE MANY INDISPUTABLE FACTS OF NATURE is that wood is easier to cut than concrete. Therefore, it's important to size the timber-frame foundation to accommodate the brick pavers you'll be using for the shed floor.

Start by laying out a long row of the bricks on a flat surface, such as a driveway or garage floor. Press the bricks together to form a row that's approximately 72 in. long. Lay a second row of bricks perpendicular to the first one until it measures about 48 in. (as shown). Remember, the floor frame doesn't have to be exactly 4 ft. wide by 6 ft. long, but it must be able to accept a whole number of bricks.

Once you have the proper number of bricks laid out, measure each row and add ¼ in. for clearance; these measurements represent the inside dimension of the frame. To obtain the overall size of the frame, add 7 in. to each dimension (this is to accommodate two 4×6s, which each measure 3½ in. wide). Double-check your measurements and math, then cut to length the four 4×6 timbers.

LEAN-TO SHED LOCKER

This drawing reveals the construction method used to build the Lean-to Shed Locker. Note the timber-frame foundation, brick-paver floor, and plywood siding.

Metal flashing

½-in.-thick plywood roof sheathing

Faux-slate shingles

1×4 ledger

Thirty-degree roof slope

1×4 fascia

2×4 rafter

14¼ in.

1×4 corner board

Header

2×4 top plate

Grooved plywood siding

8-ft.

Plywood door with 1×4 face frame

Jack stud

2×4 sole plate

Brick pavers

24 in.

4×6 timber-frame foundation

Approximately 4 ft.

Approximately 6 ft.

Crushed-granite base

Half-lap corner joint

SAFETY FIRST

Be sure to wear safety glasses, gloves, and a dust mask when sawing pressure-treated wood. Toxic chemicals are used to treat the lumber, and you don't want to inhale any of the dust. In fact, you should wear safety gear when sawing any type of lumber, because wood dust and chips of any kind can irritate your lungs and injure your eyes.

PRO TIP

Before sawing a half-lap joint, mark the waste area with an "X" to make sure you don't cut on the wrong side of the line.

TRADE SECRET

The wood preservatives used in pressure-treated lumber don't soak all the way into the middle of the board. Therefore, after cutting half-lap joints in 4×6 timbers, apply a generous coat of wood preservative to the freshly cut ends. Use a wide paintbrush and spread the preservative across the joint's cheek and shoulder cuts and on the end of the timber.

Circular Saw Safety Tips

A PORTABLE CIRCULAR SAW IS AN indispensable shed-building tool, but like all power tools, it's potentially dangerous. Heed the following precautions when using this popular power saw:

- Never operate a saw that has a frayed or cracked power cord.
- Check to make sure the lower blade guard glides up smoothly and retracts automatically.
- Immediately replace a dull or damaged blade.
- Always wear eye and ear protection.
- Use moderate pressure to slowly guide the saw through the cut.

- Never force the tool; let the blade do the cutting.
- Clean the board of all dirt, grease, and other debris before cutting.
- Before squeezing the trigger switch, make sure the sawblade isn't touching the edge of the board.
- Keep both hands on the tool while sawing; never hold the board across your knee and cut with one hand.
- Unplug the saw before adjusting the depth of cut or changing the blade.
- Never operate any power tool when you are fatigued or under the influence of alcohol or medication.

I used a portable circular saw to cut the four timbers to the proper length. A 7¼-in. circular saw has a maximum cutting capacity of about 2¼ in.; a larger, 8¼-in. model will cut approximately ⅝ in. deeper. To cut clean through the 3½-in.-thick timber, you'll need to make one cut, flip over the timber, then make a second pass from the other side.

After trimming the four frame pieces to length, set them on sawhorses and prepare to cut a 2¾-in.-deep by 3½-in.-wide half-lap joint in each end of the four 4×6s.

1. Instead of cutting each timber individually, clamp them together with their ends flush, measure 3½ in. from the end, and use a framing square to mark a square line across them. After adjusting the circular saw to its maximum depth of cut, make a shoulder cut along the pencil line, as shown in **A**.

2. Then make a series of closely spaced cuts through the waste-wood area, being careful not to saw beyond the initial shoulder cut, as shown in **B**. Remember that the circular saw will cut only about 2¼ in. deep, but the half-lap joint needs to be 2¾ in. deep. Therefore, use a handsaw to extend each circular-saw kerf to the bottom of the joint.

5. Check the frame for square by measuring the diagonals. When the two dimensions are identical, the frame is square. Fasten the 4×6s together by driving two 4-in.-long landscaping screws through each half-lap corner joint, as shown in **E**. You could join the timbers with galvanized lag screws or landscaping spikes, but the hex-head landscaping screws are much quicker and easier to install. After checking to make sure the timber frame is still square, screw a 1×4 diagonally across it. This temporary brace prevents the frame from getting knocked out of square.

Prepare the site

It doesn't take long to prepare the site for a shed foundation that measures only 4 ft. by 6 ft., but that doesn't mean you can rush through the steps. To ensure that the timber frame provides long-lasting support for the shed and brick floor, you must thoroughly complete each step before moving on to the next one.

1. Mark the outline of the timber frame on the ground where you plan to construct the shed. There are two simple ways to do this. You can measure the frame, then use a hoe or shovel

3. Once that's done, use a hammer to knock away all the waste-wood slices, then use a wide-blade wood chisel to scrape the bottoms of the joints smooth, as shown in **C**.

4. Now remove the clamps and assemble the four timbers to form a rectangular frame. After treating the cut ends with preservative, overlap the 4×6s at the corners, making sure the half-lap joints fit together tightly, as shown in **D**.

TRADE SECRET

After compacting the stone base, check it for level. Add stone to any low spots and scrape off a little stone from the high areas until all four sides of the base are perfectly level. Taking the time now to level the stone base will make it much easier to level the timber-frame foundation. If you do add stone to a low area, be sure to compact it with a hand tamper to prevent it from settling.

to scratch the outline into the ground, or set the frame in position and sprinkle white flour around the perimeter to mark the outline.

2. Once the frame's outline is clearly marked on the ground, use a flat shovel to carefully remove grass from the area. Lay the strips of sod off to the side in a shady spot and water them periodically; you may need them later to fill in bare spots after the timber frame is installed. Next, use a hoe to excavate a 4-in.-deep by 8-in.-wide trench around all four sides of the outline. Pile the soil from the trench in the middle of the excavated area. Smooth out the bottom of the trench, then check it with a 4-ft. level. It doesn't have to be perfectly level, but it should be relatively close, say, within 1 in. or less across the length of the trench. Add soil to any low areas and scrape away the high spots to level all four sides of the trench.

3. The next step is to create a solid base out of crushed stone. You can use several types of aggregate, including processed stone or plain gravel, but my favorite material for this job is crushed granite. It compacts tighter and forms a much stronger, more stable base than other aggregates do. Don't use sand for the base. It's not stable enough and will eventually settle or wash out from under the timber frame.

Pour about a 2-in.-thick layer of crushed granite into the trench, as shown in **F**. Smooth out the stone with a hoe, then spray it with a garden hose. I find that the water helps the crushed stone compact more readily and mixes with the stone dust to create a cement-like mortar.

4. Use a hand tamper to pound the stone flat, as shown in **G**. You can buy a cast-iron tamper

at a hardware store, but a homemade version works just as well. To make one, take a 36-in.- to 42-in.-long 4×4 and screw a 10-in.-long 2×6 block onto one end. Center the block on the 4×4 and attach it with four 3-in.-long decking screws.

Install the timber-frame foundation

The next step is to carefully set the frame on top of the stone base. To determine its final resting spot, check whether it's aligned with the siding on the house. I do this by holding a 4-ft. level (or long, straight-edged board) against the siding that reaches the top of the timber frame. Then I move the frame in or out until its back edge aligns with the level, which typically places the frame about 1 in. to 1½ in. away from the foundation wall. Later, when the shed's wall framing is attached to the house, it'll land flush with the back edge of the frame.

1. Use a 4-ft. level to check all four sides of the frame, as shown in **H**. Add or remove some stone from underneath to make it level. It's a good idea to press down on each corner to make sure the frame isn't rocking on any high spots. If you see any gaps below the frame, fill them with crushed granite. When you're happy with the frame's position, anchor it to the ground with long pins cut from ½-in.-dia. metal reinforcing bars (rebar).

2. Using a spade bit fitted onto an extension shaft, bore ½-in.-dia. holes for the rebar near each corner of the frame. Use a hacksaw or sabersaw equipped with a metal-cutting blade to cut four 18-in.-long pieces of ½-in.-dia. (No. 4) rebar. Then drive the rebar through the hole and into the ground with a small sledgehammer or an engineer's hammer, as shown in **I**. Tap the pins flush with the top of the 4×6, but be careful not to smash the frame with the hammer; otherwise, you may knock it out of level. After installing all four rebar pins, you can remove the 1×4 diagonal brace.

Lay the brick floor

The last stage of building the timber-frame foundation is laying the brick floor. I like to use concrete pavers for floors rather than traditional clay bricks; concrete pavers are much harder and less likely to crack. For this project, I used a combination of dark red and charcoal gray pavers, each one measuring 2⅜ in. thick by 3⅞ in. wide by 7⅞ in. long, and installed them in a basket-weave pattern.

PRO TIP

Invest in a high-quality 4-ft. level. Bargain-bin levels give inaccurate readings, making it impossible to build level foundations and plumb walls.

PRO TIP

A timber-frame foundation will accept a variety of flooring materials, including marble chips, gravel, and concrete, which can be colored with masonry dye.

WHAT CAN GO WRONG

If you live in an area that receives below-freezing temperatures, a brick-paver floor may be subjected to occasional frost heave. In extreme cases, the bricks can be pushed up enough to block the doors from swinging open. If that occurs, trim an extra ½ in. off the bottoms of the doors. By the way, once the ground thaws, the bricks should settle back down. If they don't, tap them down with a rubber mallet.

TRADE SECRET

After using a screed to scrape the stone base flat, you'll end up with a pile of stone along the front edge of the foundation. The easiest way to remove this excess stone without disturbing the compacted stone base is with a rectangular steel trowel. Simply slide the trowel under the excess stone, then lift it up and dump the stone into a bucket.

1. The first step is to fill with crushed granite the area inside the frame to within 2 in. of the top. However, if you have 3 in. or 4 in. to fill, you can first add a couple inches of soil so you won't need as much granite. Be sure to pick out any rocks, roots, or other debris, then rake the soil smooth before compacting it with a hand tamper. Be careful not to hit the timber frame with the tamper.

2. Next, dump a few bags of crushed granite onto the compacted soil, as shown in **J**. Then, smooth out the gravel using a garden rake.

3. Thoroughly compact the stone, as shown in **K**. Spend at least three minutes pounding the surface flat, add a little more stone, then compact the surface again. Repeat this process until the stone is within 2 in. of the top of the frame, then add a final 1 in. or so of loose stone.

4. I used a screed to level the stone. To make one, cut a 2×3 or 2×4 about 3 in. longer than the length of the timber frame, then cut a 1×4 about 2 in. shorter than the inside dimension of the frame. Screw the 1×4 to the 2× board so that it hangs 2 in. below the 2×. To use the screed, set it on top of the timber frame and pull it toward you so the 1×4 scrapes away the excess stone and leaves a flat, level surface, as shown in **L**. If the screed skips over a low spot, fill the depression with stone, then smooth it with the screed again.

Compact the surface one more time, add a little more crushed granite, then strike it off with the screed. Finally, use a garden hose to spray a fine mist of water across the stone. The surface is now ready for the brick pavers.

in . If a paver chips or cracks in half, replace it. Continue setting the brick pavers in this manner until the frame is filled. If the 4×6 timbers were sized and cut correctly, the last row of bricks will fit easily within the frame. If they don't fit, you'll have to cut them to size with either a motorized wet saw (available at tool rental shops) or a portable circular saw fitted with a masonry blade.

6. Once all the pavers are set in place, pour a thick layer of fine-grain play sand onto the surface. Use a broom to sweep back and forth across the surface several times to drive the sand down between the pavers, as shown in . The sand will fill any voids and help lock the pavers together. The excess sand should be swept up and saved. In about three or four months, after the original sand has settled or been washed out by rain, use the extra to fill in the joints again.

Wall Framing

The walls and roof of this shed are framed entirely with pressure-treated 2×4s. You can use standard untreated 2×4s for much of the framing, but treated lumber offers greater resistance to water damage and wood-munching insects. However, you absolutely must use pressure-treated lumber for the sole (bottom) wall plates, the jack studs, and the rear corner studs that are attached to the house. Those parts are most susceptible to moisture damage.

Build the end walls

Start by building the shed's two 24-in.-wide end walls. For each wall, cut two 76¾-in.-tall wall studs and two 24-in.-long plates, one of each for the top and the bottom of the wall.

5. To set the bricks, start in one corner and work out in both directions. To create the basket-weave pattern, set the bricks in alternating right-angle pairs: Set two pavers parallel to the house, followed by two pavers perpendicular to it. Use a rubber mallet to pound each paver flush with the top of the frame, as shown

1. I like to work at waist level, so I lay the studs across sawhorses before attaching the top plates and sole plates with 3-in. decking screws, as shown in **A**. Note that it is much easier to drive screws when you drill pilot holes first.

2. Next, stand one end wall on top of the timber-frame foundation and press its rear stud flat against the house. Align the outer edge of its sole plate flush with the outside of the 4×6 frame. Then fasten the wall in place by driving three 3-in.-long decking screws through the sole plate and into the frame, as shown in **B**. Place one screw near each end of the sole plate and one in the middle.

3. Using a 4-ft. level, make sure the 2×4 frame is plumb—that is, perfectly vertical—in two directions: from side to side and from front to back. Then secure the frame to the house wall with four 3-in. screws, as shown in **C**. Place one screw near the top and bottom of the rear stud and equally space the other two screws in between. Repeat this process to install the second end wall.

Build the front walls

Once the two end walls are in place, it's time to construct the two narrow front walls, which are only 14¼ in. wide.

D

F

E

walls create a 40-in.-wide doorway. If you need a wider opening, make the front walls narrower.

2. Next, stand a front wall up against—and perpendicular to—one of the end walls you installed earlier. Although it's not necessary, I like to clamp the two walls together while I make sure the outside edges are flush and the tops of the walls are even (it may be necessary to slip shims under the front wall). When the walls are aligned, screw them together with four 3-in. screws, as shown in **E**.

3. Unlike with the end walls (which sit on the timber frame), you can't drive screws through the sole plate of the front wall because it sits on top of concrete-brick pavers. Therefore, you must install a diagonal 2×4 brace to anchor the front wall. Use a framing square to make sure the two walls form a 90-degree angle, then screw a 23-in.-long 2×4 brace diagonally across the two sole plates, as shown in **F**. The brace will add rigidity to the front wall and eliminate any wobbling that may occur when you open or close the shed doors.

1. Follow the same steps as for the side walls: For each wall, cut two 76¾-in.-tall wall studs and two 14¼-in.-long plates, then place the studs on sawhorses and screw on the top plates and sole plates, as shown in **D**. Note that the width of the two front walls affects the size of the doorway opening. In this case, the 14¾-in.-

Frame Out the Doorway

The next step is to finish framing the doorway opening by installing the jack studs and the header.

1. Cut two 2×4 jacks (sometimes called trimmer studs) to 76¼ in., and install one jack stud on the end of each front wall, as shown in **G**. Check each jack for plumb, slipping shims, as necessary, between the jack stud and the wall stud to keep it straight and plumb, then fasten the jacks with 2½-in. decking screws. Note that the jacks are cut 3½ in. shorter than the front walls to accommodate the header.

2. To make the header, first measure the distance between the two front walls. Then cut two 2×4s to that dimension and screw together the two boards to create the header. Set the header on edge on top of the jack studs, as shown in **H**. Check the header with a 4-ft. level and, if necessary, shim up one end to make it perfectly level. Then screw each end of the header to the front walls.

3. On this shed, I installed a perforated hardboard panel (commonly known as pegboard) on the wall of the house over the siding. If you decide to install one of these tool-storage panels, you'll need to install the 48-in.-wide by 73-in.-long sheet now, because it won't fit through the doorway once the plywood siding is installed. First, though, mount two horizontal 2×4s to the back wall. Space them 48 in. apart and screw them to the end walls. Then slide the ¼-in.-thick perforated panel into position and fasten it to the 2×4s with 1¼-in. screws, as shown in **I**.

Roof Framing

The shed-style roof on this structure has a slope of 30 degrees, which is the perfect angle for this particular size building. A shallower slope wouldn't shed rain or snow as well and a steeper slope would significantly reduce the height of the doorway opening.

The roof frame consists of four 2×4 rafters and a 1×4 ledger board. To simplify the installation process, the rafters are first fastened to the ledger, then the entire assembly is lifted onto the shed walls. This prefabricated building method is not only faster and more accurate than the piece-meal approach, but it is also safer because you spend much less time working on a ladder.

Cut the roof rafters

The first step in building the roof is to cut one rafter to length and check its fit by placing it in position on top of the wall. This initial rafter is used as a template for marking the remaining three rafters. It's important that all four rafters are exactly the same size. If they're not, you'll have trouble getting the plywood sheathing and roof shingles to fit properly.

Because this roof has a slope of 30 degrees (or a pitch of slightly less than 7-in-12), both ends of each 2×4 rafter are miter cut to a 30-degree angle. Also, the bottom corner of each rafter is lopped off to make it sit flat on the top wall plate. Although you can determine the length and

Making a Roof-Rafter Jig

In carpentry, there are many times when the most precise measuring method doesn't require a measuring tape. For example, determining the exact length of a roof rafter for this shed roof is best done with a simple wooden jig. The jig consists of two 18-in.-long pieces of 1×4 and one 8-in.-long 1×4 block. The longer pieces represent the 2×4 rafter and the short block acts as the ledger board. Cut a 30-degree miter in one end of one 18-in.-long board. Screw the 8-in. block to the mitered end of the 1×4. This piece represents the top half of the rafter with the ledger attached.

Make an identical 30-degree miter cut on one end of the second 18-in.-long board. Measure 3⅛ in. along the mitered end and make a mark, as shown in the drawing below. Hold a combination square against the mitered end and scribe a line across the face of the board at a right angle to the 30-degree miter. Cut along the line to remove the corner from the board. This piece represents the bottom half of the rafter.

To use the jig, hold the two halves together to form a mock rafter. Set it on top of the shed's end wall, with the 8-in.-long ledger block pressed tightly against the siding. Adjust the length of the rafter by sliding the two halves closer together or further apart. When the front edge of the rafter is flush with the front edge of the top wall plate, draw a pencil line along the seam between the two halves, as shown.

Screw together the two halves of the jig, using the pencil mark as an alignment guide. Position the jig on the wall again to make sure it fits properly. Unscrew the 8-in. ledger block and use the jig as a template to mark cut lines on a length of 2×4. Cut one rafter and test-fit it on top of the wall; don't forget to slip the 1×4 ledger block behind the rafter. If it fits, mark and cut the remaining rafters.

A simple rafter jig can be used to help determine rafter length.

This simple jig provides an easy way to determine the length of the 2×4 roof rafters.

30-degree angle

1 × 4 × 1½ ft.

1 × 4 × 1½ ft.

30-degree angle

3⅛ in.

90-degree angle

Cut-off bottom corner

1 × 4 × 8 in. ledger block

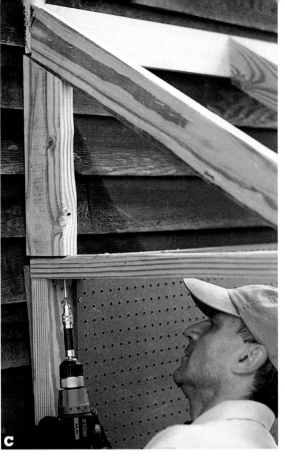

If you mistakenly cut a rafter a little too short, don't toss it onto the scrap heap. You may be able to save it. Cut a thin slice of wood equal in thickness to the amount you undercut the rafter. In other words, if the rafter is ¼-in. short, cut a ¼-in.-thick shim. Glue the shim to the ledger board, hold the undersize rafter against the shim, then screw it in place.

PROTIP

Use a sharp, thin-kerf, carbide-tipped blade in your portable circular saw. It'll cut quickly and cleanly with less risk of dangerous kickback.

layout of the rafters mathematically, it is simpler and more accurate to do it with a homemade rafter jig made from three scrap pieces of 1×4. (For more details, see Making a Roof-Rafter Jig on p. 89).

The quickest, most accurate way to execute these angled cuts is with a power miter saw. You can also use a portable circular saw; just be sure to clamp the 2×4 to a sawhorse so you can keep both hands on the tool. Use a framing square or sliding bevel square set at a 30-degree angle to accurately mark the cut lines.

Assemble the roof frame

After making all four rafters, cut the 1×4 ledger board to length. The ledger should be the same length as the timber-frame foundation, which is also equal to the width of the shed, as measured from one end wall to the other. The ledger on this shed measures 78⅜ in.

1. To build the roof assembly, first clamp the ledger board to sawhorses, then mark layout lines for the four 2×4 rafters. Rafters are commonly spaced on 24-in. centers; however, on this simple roof, I positioned one rafter flush at each end of the ledger and equally spaced the other two in between, as shown in **A**. Attach each rafter to the ledger by driving two 1⅝-in. decking screws through the ledger and into the mitered end of the rafter.

2. The next step is to set the rafter assembly on top of the shed walls, pressing the ledger flat against the siding and checking to make sure the rafter ends are flush with the edge of the front wall. Be sure to align one end of the ledger board with the outside edge of the end wall by holding a 4-ft.level against the end wall and allowing it to extend past the ledger. Then drive one 2½-in. decking screw through the ledger and into the wall of the house. Move to the other end of the ledger, place a 4-ft. level on top, raise or lower the ledger until it's perfectly level, then screw it to the house, as shown in **B**. Secure the ledger with 10 more screws, making sure you catch at least two studs inside the house wall. If you screw into only the sheathing behind the siding, the ledger may pull away from the house over time.

3. With the ledger in place, screw the front end of each rafter to the top of the wall with 2½-in. screws. Finish up the roof framing by measuring and cutting a short vertical 2×4 to fit against the house wall, directly beneath the outside rafters. These pieces of blocking, which are mitered at a 30-degree angle at one end, will provide solid support for the plywood siding. To fasten the 2×4 blocking, drive a screw up through the top wall plate from below, as shown in **C**. Then drive another screw at an angle down through the inside of the rafter and into the top of the 2×4 blocking.

Siding and Trim

The exterior of this shed is covered with ⅝-in.-thick textured plywood siding. Its roughsawn surface has vertical grooves spaced 4 in. apart. I chose plywood siding because it goes up quickly and adds strength and rock-solid rigidity to the 2×4 framing. You'll need three 4-ft. by 8-ft.

plywood sheets for this shed: two sheets for the siding and one sheet for the doors.

All the exterior trim is cut from roughsawn red cedar 1×4s. I like using cedar for the trimwork because it takes finish well, it's naturally rot resistant, and it's dimensionally stable, so it's not likely to twist, warp, or spilt. Despite the shed's compact dimensions, there's quite a bit of cedar trim—nearly 80 linear ft. of trim is installed along the edges of the roof, at all the corners, and around the doorway opening. In addition, there's another 30 linear ft. of trim on the face frames of the two swinging doors.

Install the siding

Before I cut any siding or trim, I generally stain or paint it first. Besides being easier to do ahead of time, this helps protect the wood better.

1. Rip one of the full 4-ft. by 8-ft. sheets of plywood siding in half for the 24-in.-wide end walls (ripping means to cut lengthwise or along the grain, not across it). These pieces won't be exactly 24 in. wide because a little wood is lost to the saw kerf, but the cedar corner boards will hide any gaps.

2. Cut the upper ends of the two plywood end pieces at a 30-degree angle to match the roof slope. Check the fit, making sure the siding extends from the bottom of the sole plate to the top of the rafter. If the panels are a little short, that's fine. The rake trim will cover any spaces. However, the siding mustn't stick up above the rafters, or it will interfere with the installation of the roof sheathing. When the panel is flush with the wall's sole plate, pressed against the end wall, and tight against the house siding, secure it with 1½-in. ring-shank siding nails spaced about 10 in. apart, as shown in **A** on p. 92.

A

3. After the two sides have been nailed in place, measure the two front walls and cut the plywood siding to fit. Note that the pieces must be wide enough to overlap the jack studs and tall enough to cover the ends of the rafters. Fasten them to the framing with siding nails spaced on 10-in. centers, as shown in **B**.

Finally, cut a narrow spacer strip of plywood siding to fit across the header and nail it in place. It doesn't matter whether the grooves on this piece run vertically or horizontally, because the entire strip will eventually be hidden by the cedar fascia board and the doorway header trim.

4. The next step is to cover the roof rafters with ½-in. exterior-grade plywood (either BCX or CDX grade), which serves as a substrate for the roof shingles. Measure and cut the sheathing to fit flush with the siding on all three sides of the shed, then slide the plywood panel into place, making sure it overlaps the ledger board at the rear, as shown in **C**. Use 2½-in. (8d) galvanized nails to fasten the sheathing to the rafters. Space the nails about 8 in. apart.

Add the trimwork

I've found that it's best to start the trimwork near the roof, then work your way down, saving the doorway trim for last. On this shed, all the trim was cut from red cedar 1×4s and fastened with 1½-in.-long galvanized ring-shank siding nails.

The first pieces to be installed are the angled rake boards that fit along the tops of the end walls. These two pieces follow the roof slope, so you must miter-cut both ends of each board at a 30-degree angle.

B

C

1. Cut the rakes to extend down from the house siding and flush with the plywood siding nailed to the front wall. Hold the top edge of each rake even with—or slightly below—the roof sheathing, then fasten it with six 1½-in.-long siding nails, as shown in **D**.

2. Measure, cut, and install the fascia board, which spans across the front of the shed from one rake board to the other, as shown in **E**. Before nailing the fascia in place, make sure its bottom edge is even with the bottom of the rake boards.

3. Now install the corner boards. Trim the two front corners of the shed with two 1×4 boards. Fit one board under the rake and nail it to the end wall; install the other one under the fascia and nail it to the front wall. The two boards overlap to form an L-shaped corner cap. I install the end wall board first so that the second corner board will overlap it and hide the seam from the front.

4. To follow the pitch of the roof, you'll need to cut a 30-degree miter on one end of an 84-in.-long 1×4. That's a few inches longer than necessary, but it allows you to cut the board to its exact length without having to measure. Hold the corner board in position under the rake board, making sure it's flush with the front edge of the rake. Have a helper make a mark where the corner board intersects the bottom edge of the plywood siding. Then cut the board to length at that mark and secure it with ten 1½-in.-long siding nails, as shown in **F**. Place two nails near each end and equally space the other six in between.

5. Measure and square-cut both ends of the second corner board to fit between the timber-frame foundation and the fascia. Position this piece so it overlaps the first corner board, then secure it to the front wall with ten 1½-in.-long siding nails.

6. The final three pieces of 1×4 trim go around the doorway opening. Start with the horizontal head trim that fits between the fascia and the top of the doorway, cutting it exactly 10 in. longer than the width of the doorway opening. The extra length allows the head trim to extend 1½ in. beyond the vertical 1×4 side trim that will be installed along each side of the opening. You'll also have to rip this 1×4 head trim to fit the narrow space between the doorway and the fascia.

7. Cut the head trim flush with the plywood siding at the top of the doorway opening, as

shown in **G**. After the head trim has been ripped to width, hold it tightly against the bottom of the fascia, center it in the opening, then nail it in place.

8. Cut the 1×4 side trim pieces to fit underneath the head trim on each side of the doorway. Align the side trim pieces flush with the edge of the plywood siding, then attach them with eight 1½-in.-long siding nails.

Roofing

Roofing is often the most mundane and tedious part of building a storage shed, but not in this case. First, the roof of this shed locker is only about 20 sq. ft. in total area. You'll be done with the roof long before you ever get tired of nailing down shingles. Plus, it's always a little exciting when you work with an innovative new building product for the very first time.

The Dura Slate shingles used on this shed look like authentic slate but are made of a high-tech polymer that's more like hard rubber. The 12-in. by 18-in. shingles are virtually indestructible, but they are as easy to install as standard asphalt shingles. However, before you can begin the installation, you must prepare the roof for the shingles.

Prepare the roof deck

The first step is to install drip-edge flashing around the perimeter of the plywood roof sheathing. The aluminum flashing, which is sometimes called drip cap, prevents water run-off from seeping behind the siding. I installed white drip edge because it matched the white-stained trim. It's also commonly available in brown and unpainted aluminum, which is known as a mill finish.

ously installed flashing. Be sure to position the flashing so it overlaps the flashing on the front edge of the roof, as shown in **B**. Then nail the flashing to the plywood sheathing. Use ¾-in. roofing nails spaced 10 in. apart; drive the nails about 1 in. from the rear edge of the flashing.

Install the shingles

Before the main shingling can begin, you must install a starter course of Dura Slate shingles along the three edges of the roof. This initial layer provides support for the roofing and fills in the spaces between the shingles so you won't see the plywood sheathing.

1. Make a starter shingle by cutting in half a full-size Dura Slate shingle. The simplest way to cut the rubberlike shingle is with a sharp utility knife. Don't try to cut all the way through in one pass; the stuff is much too tough. Instead, score the surface three or four times, using a straightedge to help guide the cut, then bend the shingle until it snaps in two. Discard the bottom part embossed with the slate texture and keep the smooth top portion. Cut enough starter shingles to cover both ends and the front edge of the roof.

2. Begin installing the starter course along the sloping ends of the roof. Align the starter shingles with the edge of the drip-edge flashing and attach them with roofing nails, as shown in **C** on p. 96. Then nail the starter shingles along the front edge of the roof, again keeping them flush with the flashing. After the starter course is installed, you can begin shingling the roof.

3. Draw a vertical centerline down the middle of the roof and use it as a starting point for installing the main shingles. Position a full-size

1. Start by placing a length of flashing against the front edge of the roof. Hold one end of the flashing even with the outside surface of the rake board, then mark the other end of the flashing where it intersects the opposite rake board. Cut the flashing at the mark with a pair of tinsnips and attach it to the roof with ¾-in. roofing nails spaced 10 in. apart, as shown in **A**.

2. Install drip-edge flashing along each side of the roof. Cut the pieces so they extend from the house siding to the front edge of the previ-

PRO TIP

The best tool for cutting aluminum flashing is a sharp pair of tinsnips. If you don't own a pair, try using poultry shears or large scissors.

TRADE SECRET

After cutting faux-slate shingles to fit along the roof edge, perform a little handiwork to make the shingles' edges blend in with the rest of the roof. First, round off the square corner with a utility knife and wood rasp (a very coarse file). Then shave the shingle's just-cut edge with the utility knife to create an uneven bevel. Try to mimic the realistic texture embossed on the opposite, uncut edge.

Dura Slate shingle with its right-hand edge on the centerline. Make sure the shingle's bottom edge is even with the drip-edge flashing, then secure it to the plywood sheathing with two roofing nails, as shown in **D**. Place the nails about 1 in. in from each edge of the shingle and 9 in. up from the butt (bottom) edge. Now work out in both directions from the centerline, laying full-size shingles to the edge of the roof. When you get to the end, hold the last shingle in place and mark where it overhangs the edge, as shown in **E**.

4. To cut the last shingle, lay it on a hard, flat surface and use a straightedge to guide you as you score the shingle with a sharp utility knife, as shown in **F**. Then bend the shingle until it snaps along the scored line, slicing it from the back, if necessary, to separate the two pieces. Align the cut shingle with the bottom edge and the end of the roof and attach it with two roofing nails.

5. To begin installing the second course of shingles, move back to the roof's centerline, as shown in **G**. This time, center the first shingle directly over the centerline. This offset alignment ensures that the vertical seams between the shingles will be staggered from one course to the next. If the seams aren't staggered, water will flow between the shingles and eventually leak into the shed.

Again, work out from the center of the roof, installing full-size shingles and maintaining a 6-in. exposure to the weather. Trim the end shingles to fit, as described previously. For the third course, return to the centerline and follow the same procedure as the first course: Align the right-hand edge of the first shingle with the centerline. Continue working your way up the roof, trimming the shingles in the final course to fit within ½ in. of the house siding. This particular roof requires five full courses of shingles.

Install the flashing

A continuous piece of aluminum, lead, or copper flashing must be installed along the seam where the roof meets the house. Without the flashing, water will run down the siding and into the shed. Flashing is available in various widths; I used 8-in.-wide flashing for this shed. You'll need enough flashing to cover the entire width of the roof and ¼ in. on each end. That little bit extra will permit the flashing to overhang the shingles by ⅛ in. on each end of the roof.

Although lead flashing is very malleable, aluminum flashing must be bent lengthwise down the middle to give it a slight L-shape profile. If you use copper flashing, your metal fabricator can make the bend for you.

1. Before you can install the flashing, you must pull the nails out of the siding course directly above the shed roof. Then, using a flat bar, carefully pry the bottom edge of the siding away from the wall by about 1 in., inserting shims behind the siding to prop it away from the wall. Slip the metal flashing behind the siding and allow its lower edge to extend onto the top shingle course, as shown in **H**.

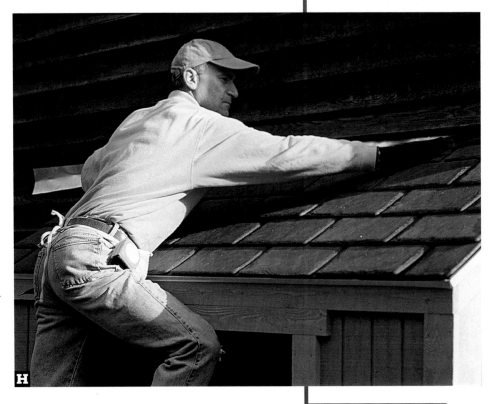

Always wear gloves when cutting and installing metal flashing. The material is very thin and bends easily, but the edges and corners can be razor sharp, especially right after they're cut. Also, never force a length of flashing into place. If your hand slips and slides along the edge, you could receive a deep, serious cut.

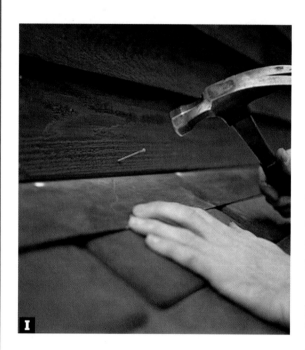

2. Remove the shims and nail the siding in place, securing the flashing at the same time, as shown in **I**. Important: Don't drive any nails through the lower flange of the flashing and into the roof.

If the house is sided with shingles, pull the nails from each individual shingle in the course directly above the ledger. Then slip the flashing behind the shingles and nail them back in place. For a house covered with vinyl or aluminum siding, the job's a bit trickier. Unhook the butt edge of the siding course immediately above the ledger from the course below it. The pros use a siding removal tool—known as a zip tool or

zipper—to disengage the butt edge, but a narrow prybar or straight-blade screwdriver works just as well.

After unhooking the siding, slip the flashing behind it. Nail a starter strip—available wherever the siding is sold—along the top of the course directly below the piece of siding you just disengaged. Use 1½-in.-long aluminum nails and space them about 10 in. apart. Drive the nails into the slots in the starter strip and through the flashing, but be careful not to drive the nails too tightly; the strip needs to be able to expand and contract freely. Finish up by hooking the loose siding course to the starter strip.

Doors

This shed features a pair of swinging plywood doors, which are the simplest, quickest type of door to build. It takes about only an hour to build both doors.

Each door consists of a grooved plywood panel trimmed with a 1×4 cedar face frame. The door panel is stained to match the shed's siding, and the face frame has the same light-gray finish as the exterior trim. For extra protection against moisture-related problems, apply a coat of stain to the back and all the edges.

I hung each door with a pair of 8-in. surface-mounted T-hinges, though I could have chosen smaller hinges. In that case, I would have used three hinges per door.

Cut the plywood door panels

Both doors can be cut from a single sheet of grooved plywood. To determine the size of each door panel, first measure the width of the doorway at the top, middle, and bottom of the opening. If everything is square and level, all the measurements should be the same; if there's any discrepancy, choose the smallest dimension. Subtract ½ in. (for clearance) from the width of the opening, then divide by two. The result is the width of each door panel.

To determine the height of the doorway, repeat the same measuring procedure, but this time take the measurements at the left side, middle, and right side of the opening. Then subtract ½ in. to find the height of the door panels. Finally, mark the door height of 75¾ in. on the plywood and crosscut the sheet with a portable circular saw. Mark the door width on the plywood and cut out the two panels, making each one 19¾ in. wide.

Attach the face frame

The face frame on each door is made from five 1×4 boards: two long vertical stiles and three short horizontal rails.

1. Lay one of the door panels across two sawhorses and cut the two long stiles to the same length as the door panel, then use a caulking gun to apply a bead of construction adhesive to the back of the 1×4 stiles, as shown in **A**. Clamp each stile to the face of the door panel with three 3-in. spring clamps, then drive 1¼-in.-long decking screws through the rear of the panel.

2. Next, measure and cut the three 1×4 rails to fit between the stiles, then glue and screw the top and bottom rails flush with the door ends. Position the middle rail so its top edge is 31¾ in. from the bottom end of the door, as shown in **B**. This places the rail slightly below the midpoint, which makes the doors appear taller. Repeat this procedure to attach the face frame to the second door panel.

Mount the hinges

Each door is installed with a pair of 8-in.-long T-hinges. Before screwing the hinges to the doors, you must first install 3½-in.-wide by 5½-in.-long hinge blocks cut from a cedar 1×4 to support the long hinge leaves. I cut fanciful spear points in the hinge blocks, but you can leave them square or cut a different shape; just make sure the blocks are at least 3½ in. by 5½ in.

PRO**TIP**

Pay extra for heavy-duty, exterior-grade door hinges. Thin-gauge metal hinges don't adequately support doors with storage racks mounted on the back.

1. Position the center of the lower hinge blocks 10 in. up from the bottom of the door and the upper hinge blocks 8 in. down from the top of the door. Attach each block with construction adhesive and three 1¼-in.-long deck-

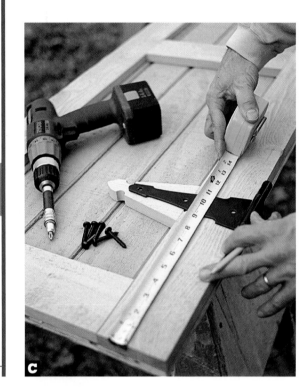

ing screws driven through the rear of the door panel. Bore ⅛-in.-dia. pilot holes to prevent the screws from splitting the hinge blocks.

2. Hold each hinge directly over the center of the hinge block and mark the screw-hole locations, as shown in **C**. Bore ³⁄₃₂-in.-dia. pilot holes through the hinge block and attach the hinge with the pan-head screws provided.

3. The right-hand door has an astragal strip attached to its edge. The purpose of this narrow strip is to hold the two doors even when they're both latched shut. Cut a 2¼-in.-wide by 75¾-in.-long astragal strip from a piece of ½-in.-thick exterior-grade plywood. Position the astragal on the inside of the right-hand door, allowing it to overhang the edge by 1 in., as shown in **D**. Then glue and screw the strip to the door.

Hang the doors

1. Place two ⅜-in.-thick shims on the brick floor in the doorway, then stand one of the doors on top of the shims and tilt it into the opening, as shown in **E**. The shims will automatically create a ⅛-in. gap at the top of the door and a ⅜-in. clearance space below it.

2. Press the hinge side of the door tightly against the doorway trim, then drive one of the supplied hinge screws through each hinge leaf, as shown in **F**. Before driving the remaining screws, remove the shims and check to make sure the door swings freely and closes without binding. If necessary, remove one of the hinge screws and tilt the door. Install the second door in the same manner. Test the swing of both doors, then install the remaining hinge screws.

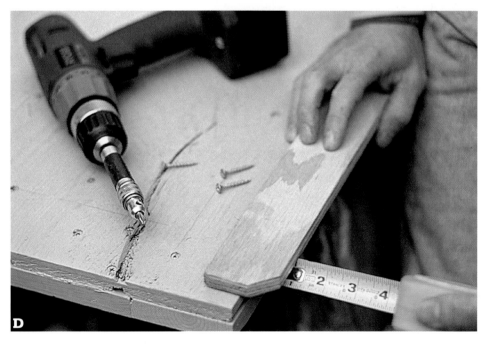

Install the hardware

The right-hand door—the one with the astragal strip—is equipped with a 3-in. sliding barrel bolt that holds the door closed. The bolt slides up into a hole bored in the header. Without this piece of hardware, both doors would sway in and out, even when latched together.

1. Position the barrel bolt against the astragal strip and flush with the top of the door, then attach it with the short screws provided. Close the door and raise the barrel bolt until it hits the header, then rap the bottom of the bolt with a hammer to leave an impression on the header.

2. Using a drill bit that's the same diameter as the barrel bolt, bore a hole through the center of the impression. Close the door and slide the bolt into the hole. If the fit is too tight, place the drill bit in the hole and wiggle it around a little to enlarge the hole.

3. Finally, measure 36 in. up from the bottom of the door and install a lockable hasp.

Interior Shelves

Inside the shed, tucked around the corner to the left and right sides of the doorway, is a surprising amount of storage space. If left as is, these niches are perfect for standing up tall items, such as pole pruners, post-hole diggers, and garden stakes.

In this storage shed, I decided to utilize the space by installing a few wooden shelves. Another option would be to put up a wire-rack shelving unit, similar to the one mounted on the back of the left-hand door.

TRADE SECRET

Every shed needs some sort of lock, if not for security reasons, then to keep out curious kids. A standard hasp and pad-lock work fine, but it's easy to misplace the lock. A better option is a lock-able hasp, which has its own keyed locking mech-anism. Screw the hinged portion of the hasp to the left-hand door. Mount the locking mechanism on the opposite door. Rotate the lock 90 degrees to hold the doors closed. To lock the hasp, use the key.

TRADE SECRET

A hasp is used to lock the two doors together, but you must also install a 3-in. sliding barrel bolt on the inside of the right-hand door to keep the doors from bulging outward. Screw the bolt flush with the top of the door. Then pull the door closed and tap the bottom of the sliding bolt with a hammer to leave an impression on the underside of the header. Open the door and bore a ⅜-in.-dia. by 1-in.-deep hole in the header; be sure to center the bit on the impression mark. Latch the door by sliding the bolt into the hole.

TO GET THE MOST OUT OF THIS COMPACT SHED locker, take advantage of every square inch of available storage space. That includes the backs of the doors, which together offer more than 18 sq. ft. of surface area. The challenge is finding the best way to utilize this area without taking up too much space inside the shed. Wire-rack shelving units offer the perfect solution. These rugged steel racks are sold at most home centers and hardware stores in a wide variety of sizes and styles.

For this shed, I bought an eight-tier wire rack measuring 18 in. wide by 72 in. long. Its 5-in.-deep shelves are just right for storing cans, bottles, and small hand tools. To fit the rack onto the back of the left-hand door, I cut off the top shelf with bolt cutters. That reduced the shelving unit to about 64 in. tall. By the way, if you don't have bolt cutters, you can trim the wire rack with a hacksaw or sabersaw fitted with a metal-cutting blade.

To mount the rack, measure 10 in. down from the top of the door and draw a level line. On the line, screw two plastic mounting hooks (provided with the racks) spaced about 14½ in. apart. Set the shelving unit's top horizontal wire on the hooks and press the rack flat against the back of the door. Check to make sure the shelving unit is centered left to right. Then screw two plastic mounting hooks to the bottommost horizontal wire, but this time place the hooks upside-down to prevent the shelving unit from lifting out of the two top hooks.

On the back of the right-hand door, I wanted to hang some larger tools and supplies, so I simply tapped in a few galvanized nails. However, I could have installed a wire rack on this door as well.

A pair of bolt cutters easily slices through this powder-coated steel-wire rack. You can also use a hacksaw or a sabersaw.

Hang the wire-rack shelving unit on the hooks. Then clip two hooks onto the bottom wire, but this time turn them upside-down.

I made these shelves from 1×12 pine, but cheaper ¾-in. plywood can be used as well. Three shelves are installed on each side of the doorway. Each one rests on a pair of 1×3 cleats mounted to the back and front walls of the shed.

1. Cut 11-in.-long cleats for the back wall, then fasten them to the corner stud and the perforated hardboard with 1⅝-in.-long screws, as shown in **A**. Make the front wall cleats 17½ in. long. That way, you'll be able to screw both ends of the cleats to a wall stud. The shelf cleats can be installed at any convenient height; I find it useful to position the lowest shelf 36 in. from the shed floor and then leave spaces of 12 in. and 16 in. between the next two shelves.

2. Measure and cut the 1×12 shelves so they span the distance between the front and rear walls. Drop the shelves down onto the cleats, as shown in **B**. You can fasten them to the cleats with screws or just leave them resting in place. There is often enough weight from the items stored on them to keep the shelves from sliding out.

DESIGN
OPTIONS
PLYWOOD SHED DESIGNS

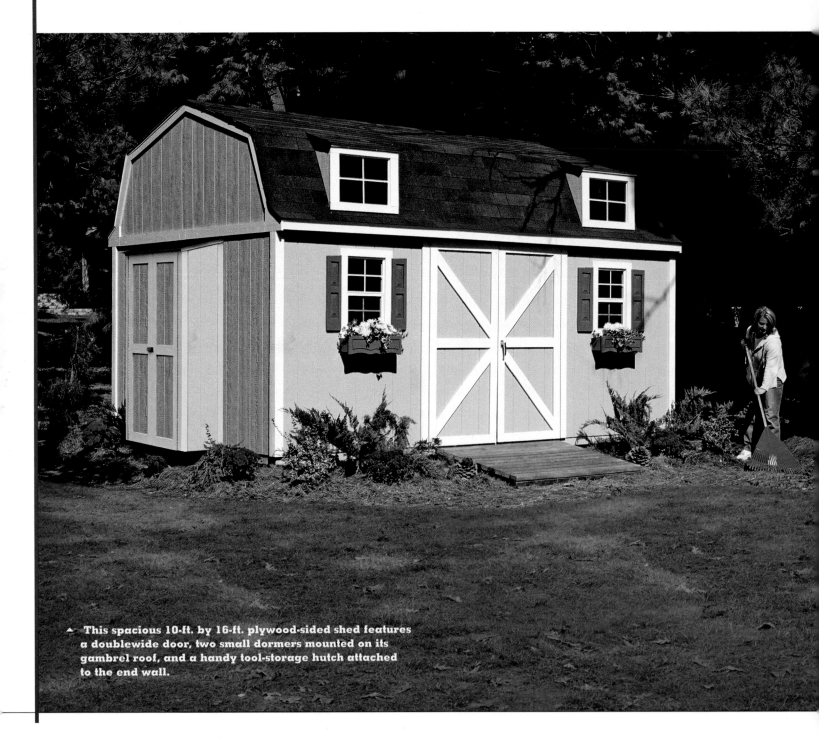

▲ **This spacious 10-ft. by 16-ft. plywood-sided shed features a doublewide door, two small dormers mounted on its gambrel roof, and a handy tool-storage hutch attached to the end wall.**

Designed for gardening enthusiasts, this 8-ft.-sq. plywood-sided potting shed has sidewalls covered with Lexan® panels. The clear plastic panels flood the interior with warm sunlight, creating the perfect environment for starting and growing plants.

This versatile 8-ft. by 14-ft. plywood-sided outbuilding can be used for general storage or as a workshop, kid's clubhouse, or artist's studio. Note how its saltbox-style roof overhangs the front façade.

Ideal for small yards, this compact 8-ft. by 10-ft. mini-barn has a gambrel roof and 4-ft.-wide doorway. Its walls are covered with rough-textured and grooved plywood siding, which is nailed directly to the wall studs.

Set against the rear of the house, this compact 6-ft. by 8-ft. storage shed is ideal for storing lawn and garden tools and a walk-behind mower. Its plywood siding comes factory-primed and ready for a topcoat paint finish.

GARDEN SHED

t 6 ft. by 6 ft., this modest outbuilding might be a bit small for general storage, but that doesn't diminish its usefulness. Despite its diminutive dimensions, this handsome Garden Shed is perfect for storing gardening tools, landscaping supplies, and lawn-care products. It'd also make a great poolside changing room or kids clubhouse.

The Garden Shed features a concrete-block foundation, 2×4 wall framing, and an attractive hip roof covered with architectural-style asphalt shingles. The batten door, which is constructed out of tongue-and-groove cedar 1×6s, is framed by two sections of picket fence.

To keep maintenance to a minimum and durability to a maximum, the exterior of the shed has fiber-cement siding, vinyl windows, and exterior trim cut from boards of cellular PVC. As a result, the siding won't rot or warp, and the windows and trim will never need painting. (To order a set of building plans for the Garden Shed, see Resources on page 214.) ▶ ▶ ▶

Solid-Block Foundation

The on-grade foundation for this 6-ft.-square shed is made of solid-concrete blocks placed at each corner of the floor frame. The blocks measure 4 in. thick by 8 in. wide by 16 in. long. Because this building site was sloping slightly, three blocks were stacked to support the rear of the shed, and two blocks were placed at the front corners. You can also use 2-in.-thick concrete blocks, called patio blocks, to help level a slanting site.

The wood floor frame measures 72 in. square, so position the blocks 70 in. apart, as measured across their outside edges. That way, the floor frame will overlap each stack of blocks by 1 in., providing clearance for the plywood sheathing to extend pass the blocks.

Set the concrete blocks

Arrange the blocks on the ground in the proper position. Next, use a flat-blade shovel to remove the grass from beneath one of the blocks. Thoroughly compact the soil with a hand tamper, then set the block back into place.

1. Check the block for level in two directions, across its width and length. If it's not level, don't hit or stomp on the block; it might crack. Instead, lay a short 2×6 on top of the block and tap on the 2×6 with a small sledgehammer, as shown in **A**. Repeat this procedure for the remaining three corner blocks.

2. Once the four corner blocks are in position, stack a second 4-in.-thick block on top, creating four two-block stacks. Now, lay a long, straight 2×4 across two stacks of blocks and place a 4-ft. level on top of the 2×4. Check for level across all of the blocks in both directions, as shown in **B**. If necessary, slip shims under the 2×4.

Floor Framing

The shed's floor frame is made entirely out of 2×6 pressure-treated lumber. The rest of the shed, including the walls and roof, are framed with untreated construction-grade lumber. However, it's important to use rot-resistant treated wood for the floor frame because of its proximity to the ground.

Center post

Copper cap

2×6 intermediate roof rafter

Ridge

2×6 roof rafter

Shingles (typically with 5½-in. exposure)

2×4 plate

¾-in. plywood

Gusset plate

Metal drip edge

2×8 fascia

2×6 frieze board

Rosette

Plywood wall sheathing typically covered with Tyvek®

2×4 wall stud

¾-in. tongue-and-groove plywood

2×6 mudsill

Concrete blocks lay 1 in. inside flooring on all sides

2×6 joist

72 in.

Corner board

Construct the mudsill

Mudsills are the bottommost pieces of wood in a building; they rest right on top of the foundation.

1. Start by cutting four 2×6s to 6 ft. long.

2. Screw two of the 2×6s together with 3-in.-long galvanized decking screws to form an L-shaped assembly, as shown in **A**. Space the screws about 12 in. apart. Fasten together the remaining two 2×6s in the same manner.

Install the joists

Set the two L-shaped assemblies on top of the foundation blocks running from the front to the rear of the shed. The 2×6 laying flat on the blocks is the mudsill; the attached 2×6 that's standing on edge is called the band joist.

1. Cut six 2×6 pressure-treated floor joists to 68⅞ in. long. It's important to cut all the joists to exactly the same length and to saw both ends of each board perfectly square.

2. Set one of the floor joists between the two 2×6s band joists and hold it flush with the end

of the mudsill. Use a drill to bore two ³⁄₁₆-in.-dia. pilot holes through the band joist, as shown in **B**. The pilot holes will make it much easier to drive in the screws, but they'll also help prevent the screws from splitting the board. Fasten each end of the joist with two 3-in. decking screws.

3. Install the remaining floor joists between the band joists, spacing them 16 in. o.c., as shown in **C**. Secure each joist with 3-in.

decking screws. After the joists are installed, check the floor frame for square by measuring the diagonals.

Put down the plywood

1. Before screwing down the shed's plywood floor, affix the floor frame to the ground with four steel-cabled ground anchors. Bolt each anchor's threaded rod to the mudsill near each corner of the floor frame. Then use a steel rod and sledgehammer to drive the anchor's pointed hold-down spike into the ground.

2. Cut two pieces of ¾-in.-thick tongue-and-groove ACX plywood for the floor: one 48 in. by 72 in. and another 24 in. by 72 in. Tap the sheets together to create a tight joint, then screw them down to the joists using 2-in. decking screws spaced 10 in. to 12 in. apart, as shown in **D**.

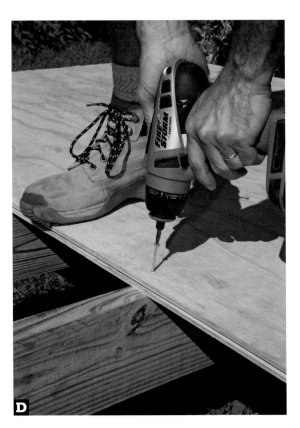

D

Wall Framing

The four walls of the shed are framed entirely out of 2×4s. Each wall is 79 in. tall, so the wall studs must be cut to 76 in. long. When the studs are set between the top and bottom wall plates, they'll create 79-in.-tall walls.

The front and rear walls are 65 in. wide, and the two sidewalls are 72 in. wide. Frame and sheathe all four walls, and erect the sidewalls first. Then set the front and rear walls between the sidewalls to create the 6-ft.-sq. building.

Note that the rear wall and both sidewalls are framed to receive a 24-in.-sq. vinyl window sash. The front wall is framed for a 29½-in.-wide batten door. You can eliminate one or two windows to create more wall-hanging storage space, but the interior of the shed will be considerably darker.

1. Cut all the parts for one of the sidewalls and lay them out on the plywood floor deck, as shown in **A**. Fasten the 2×4 wall frame together with 3-in. decking screws or 3½-in. (16d) common nails.

A

2. Once the wall frame is assembled, sheathe it with ½-in.-thick ACX plywood. Cut the plywood to extend 7 in. past the wall's bottom plate. Later, when the wall is raised, the overhanging plywood will overlap the floor frame. Fasten the plywood sheathing to the wall studs with 2-in. decking screws or 2½-in. (8d) nails spaced 10 in. to 12 in. apart.

Note that because the front and rear walls fit between the sidewalls, their plywood must extend 3½ in. beyond their end studs. That way, when the walls are erected, the overhanging plywood will cover the ends of the sidewalls.

3. After attaching the plywood sheathing, cover the wall with an air infiltration barrier, which provides an extra layer of moisture protection. Roll out the barrier, as shown in **B**, then staple it to the plywood.

Erect the walls

Don't attempt to erect any walls until you've got all four of them framed and sheathed. That way, you can use the floor deck for assembling the wall frames. Plus, there won't be any fear of one wall blowing down while you're building the others.

1. With the aid of a helper, lay one of the completed sidewalls face up on the floor deck. Then carefully raise the wall until it's standing upright on the edge of the deck, as shown in **C**.

2. Have your helper push in the bottom of the wall from the outside. Then, fasten the bottom wall plate to the floor deck with 3-in. decking screws, as shown in **D**. Space the screws about 16 in. apart.

B

C

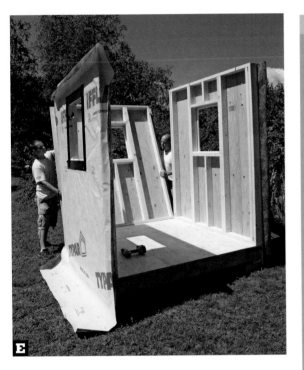

Keep It Dry

TO PROTECT THE PLYWOOD WALL sheathing from water penetration, install strips of adhesive-backed rubber flashing around the rough openings of the windows and door. Put on the bottom piece first, then the two vertical side pieces. Install the top horizontal strip last, making sure it overlaps the side strips. Now, if water seeps behind the siding, it won't soak into the plywood.

3. Once the two sidewalls are securely screwed to the floor deck, tilt up the rear wall so that it slides between the sidewalls, as shown in **E**.

4. Screw down through the bottom plate of the rear wall. Then, tie the rear wall to the sidewall by driving 3-in. screws through its end stud, as shown in **F**.

Hip-Roof Framing

The shed is crowned with a hip roof that comprises four triangular-shaped roof planes that come to a point at the peak. Hip roofs are considered one of the most complicated types of roofs to build, but don't worry. This one has been greatly simplified. There are only four rafters, which connect to a short center post at the peak. The bottom ends of the 2×6 rafters are tied together by four 2×8 fascia boards that run around the perimeter of the roof. Refer to the illustration on p. 115 for more details.

Prefab the rafters

Save yourself some trouble by pre-assembling part of the roof frame on the ground. Then have friends help lift the assembly onto the shed walls.

1. Start by cutting four 2×6 rafters to 75½ in. long, mitering each end to 30 degrees. Then double-bevel the top edge of the rafters to 25 degrees, forming two planes for supporting the plywood roof sheathing.

2. Next, cut four 3½-in.-wide by 7½-in.-long rafter mounting plates out of ¾-in. plywood, and attach one to the upper end of each rafter with 2-in. decking screws. Cut the center post from a 16-in.-long piece of 4×4. Use a compound miter saw to trim the upper end of the post, with the sawblade set for a 30-degree miter and 27-degree bevel.

3. Hold the rafter's plywood mounting plate flush with the top of the center post, then screw through the plate and into the post with 2-in. decking screws, as shown in **A**.

4. Attach all four rafters to the center post, then lift the entire assembly and place it on top of the shed, as shown in **B**. Align the lower end of each rafter directly over a wall corner.

Complete the framing

The final steps to framing the hip roof include installing a 2×8 fascia, 2×6 intermediate rafters, and 2×6 fascia. Note that all of these pieces are beveled along their upper edges to match the roof slope.

1. Cut 2×6 intermediate rafters to fit between the roof rafters. Position them 40 in. up from the fascia, so they'll support the edge of the plywood roof. Attach the 2×6s with 3-in. decking screws, as shown in **C**.

2. Once the roof frame is complete, cover it with exterior-grade plywood, as shown in **D**. You could use ½-in. plywood, but we used ¾-in. plywood so the roofing nails wouldn't poke through and be visible from inside the shed.

HIP ROOF FRAMING DETAILS

¾-in. plywood rafter mounting plate

16-in.-long center post

25⅛ in.

Intermediate roof rafter

2×6 rafter

2×6 rafter

2×6

2×6

Frieze board

6½ in.

Fascia (2×8 ripped down to 6½ in. wide)

Exterior Trim

The exterior trim installed around the windows and the door and at the wall corners looks like painted wood, but is actually a product called Azek® Trimboards, which is a cellular PVC (polyvinyl chloride). Azek cuts like wood, but will never rot, warp, or need painting.

The windows and door were trimmed with 1-in.-thick by 4-in.-wide Azek boards; the corners of the shed were fitted with prefabricated 4-in.-wide Azek cornerboards.

1. Start trimming out the windows by cutting a length of trim for the windowsill. Notch the sill to extend beyond the window opening by 2⁹⁄₁₆ in. on either side. Next, install the two vertical trim pieces, as shown in **A**.

2. Prefabricated cornerboards go up quickly and fit tightly to the building, as shown in **B**. Cut the cornerboards to extend ½ in. to ¾ in. past the bottom of the plywood wall sheathing.

Asphalt Roofing

The shed is topped with slate-gray architectural-style asphalt shingles, which are also called laminated shingles. Each 12-in.-wide by 36-in.-long shingle is composed of two strips of roofing fused or laminated together. The result is an interesting looking roof that has deep shadow lines.

Getting started

Before you can begin nailing down roof shingles, you must prep the roof by installing an aluminum drip cap.

TRADE SECRET

Rent a set of scaffolding when it's time to roof the shed. It's much safer and quicker than working from a ladder.

A

B

1. Lay the drip cap along the lower edge of the roof, making sure it overhangs the edge of the plywood sheathing, then fasten it with 1-in. roofing nails spaced 12 in. to 16 in. apart.

2. The first course of shingles is made of a starter strip, which is nailed down over the aluminum drip cap, as shown in **A**. This initial course adds extra protection to the roof edge and covers the drip cap so it won't be visible between the seams of the architectural shingles.

3. Set the first course of architectural shingles flush with the bottom edge of the starter strip. Fasten each shingle with four 1-in. roofing nails, as shown in **B**.

Completing the shingling

With the starter strip and first course of shingles in place, continue laying shingles up toward the peak. Complete one roof plane at a time.

1. Nail down the subsequent courses of roof shingles, making sure to maintain a consistent 5 in. of each shingle exposed to the weather, as shown in **C**. Also, stagger all vertical seams between shingles by at least 6 in. Trim the ends of the overhanging shingles flush with the ridge using a hook-blade utility knife. Shingle the remaining three roof planes in the same manner.

C

Whether you hand-nail the shingles or use a pneumatic roofing nailer, it's important to drive the nails straight in and just deep enough so that the heads contact the shingles. If you drive the nails in at an angle or in too deeply, the nail heads will rip the shingles, which will make the roof much more likely to suffer wind damage.

PROTIP

Most roof-shingle manufacturers recommend nailing each shingle 1 in. from each end and then 12 in. in between. Running lengthwise across each shingle is a thin stripe, called a fastening line. The roofing nails must be driven through this line to provide maximum hold-down strength and ensure that the large nail heads will be covered by the next shingle course.

D

A

2. After trimming all the shingles flush with the four ridges, cover the seams with ridge shingles, as shown in **D**. Be sure to maintain the same 5-in. exposure as you did with the architectural shingles. The best way to finish off the peak is with a copper or aluminum cap, which can be fabricated by a sheet-metal worker.

Fiber-Cement Siding

Fiber-cement siding is made of sand, cement, and cellulose fibers, but you'd never know it by looking at it. The $5/16$-in.-thick by $6\frac{1}{8}$-in.-wide siding has an embossed wood-grain surface that resembles real wood. Fiber-cement siding is ideal for outbuildings because it won't rot or crack, and it holds paint beautifully. It's also highly resistant to moisture, fire, and termites. However, its dust can be irritating. Always wear a dust mask or respirator—and, of course, eye protection—when cutting fiber-cement siding.

1. The first step is to install a starter strip along the very bottom of each wall. This narrow plastic strip bumps out the first siding course to the same angle as the subsequent courses. Fasten the starter strip to the plywood sheathing with 2-in. galvanized nails spaced 12 in. apart, as shown in **A**.

2. Measure the distance between the two cornerboards. Cut a piece of siding ¼ in. shorter than the measurement so there will be a ⅛-in. gap at each cornerboard. Hold the siding flush with the bottom of the cornerboards and secure it with 2-in. galvanized nails spaced 8 in. to 12 in. apart, as shown in **B** on p. 120.

Making the Cut

Fiber-cement siding is pretty tough stuff, and cutting it is no picnic. You could use a portable circular saw equipped with a carbide-tipped blade, as shown in the top photo below, which cuts fast but produces quite a bit of dust. If you use one, work outside and well away from people and open house windows.

Another, neater option is to use a pair of electric shears, as shown in the bottom photo below. The powerful shears cut through the siding cleanly, but making perfectly square cuts takes some practice. Electric shears can be rented at most tool rental shops. Regardless of the cutting tool used, be sure to wear a dust mask or respirator.

Portable circular saw equipped with a carbide-tipped blade.

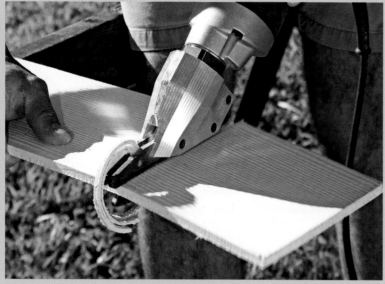

A pair of electric shears.

3. Continue to install siding up the wall, leaving 5 in. exposed to the weather. Cut the siding to fit around the windows, as shown in **C**. Position the nails ½ in. down from the top edge of the siding so they'll be hidden by the next course.

4. Once the walls are sided, fill the gaps between the cornerboards and the siding with acrylic latex caulk, as shown in **D**. Smooth out the caulk bead with a wet finger.

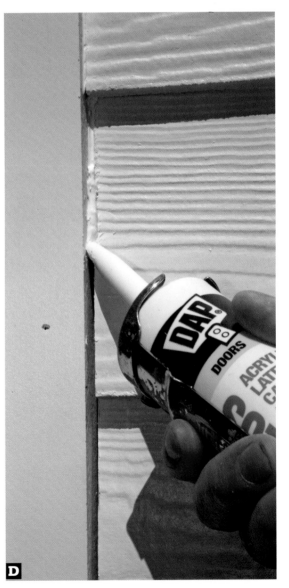

Batten Door Details

THE SHED DOOR IS ASSEMBLED FROM SIX tongue-and-groove, V-jointed cedar 1×6s. The 68-in.-long boards are held together by wood strips, called battens, which are glued and screwed to the back side. There's a horizontal batten near the top of the door, just above the 13½-in. by 16-in. decorative-glass window. The bottom half of the door is reinforced by two horizontal battens and an X-shaped batten. Note from the photo that the horizontal battens at the top and bottom of the door align with the hinges, providing 1½ in. of solid-wood fastening. The door latch is bolted to the center horizontal batten.

Note: The door comprises four full-width tongue-and-groove 1×6s. The first and last boards are ripped down to 4 ¾ in. wide.

DESIGN
OPTIONS
COMPACT SHED DESIGNS

▲ With its hexagonal shape and modest proportions, this 12-ft.-dia., six-sided structure is an attractive, unique outbuilding.

▶ At just 8 ft. by 10 ft., this handsome storage shed looks like a stable for miniature horses. It features grooved plywood siding, double-wide doors, and a cupola.

▲ This 7-ft. by 9-ft. plywood garden shed has a side hutch for storing recycling bins. Free building plans are available from Georgia Pacific.

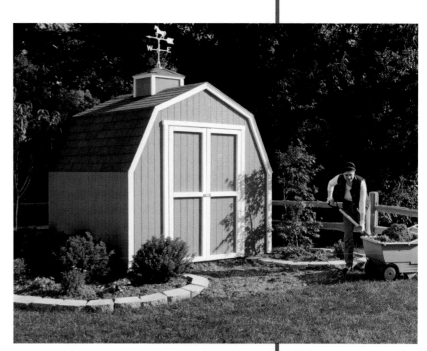

▶ Measuring just 8 ft. by 8 ft., this mini-barn will fit into the smallest yards. Its gambrel roof offers plenty of interior headroom.

▲ Nicely proportioned at 10 ft. by 10 ft., this cozy cabin features clapboard siding, a wood-shingle roof, and exposed rafter tails.

Louvered vent

Rake board

Soffit vent

Plywood floor

Roof truss

2×4 roof rafter

Ridge cap

1×4 spaced sheathing

Cedar roof shingles

2×4 wall stud

2×8 mudsill

Pressure-treated floor frame

2×6 floor joist

Plywood wall sheathing

1×4 cedar corner board

Dutch door

Platform steps

Fascia board

Double-hung window

Solid-block foundation

Bevel siding

Assembling Roof Trusses

½-in. plywood ridge gusset (both sides)

45-degree plumb cut

20 in.

1×3 stop block

100 in.

37 in.

32 in.

6¼ in.

20 in.

64 in.

1×3 stop block

20 in.

½-in. plywood side gusset (both sides)

2×4 bottom chord (ceiling joist)

2×4 rafter

Plywood floor deck

Each roof truss is built from two 2×4 rafters and a 2×4 bottom chord. The parts are assembled on the plywood floor deck. Seven 1×3 stop blocks are used to hold the three boards in position, then plywood gusset plates are glued and nailed across the joints.

A

A ROOF TRUSS IS A PREFABRICATED section of the roof frame that consists of two angled rafters and one horizontal bottom chord (ceiling joist). The reason for building the trusses now, before framing the walls, is that you can use the plywood floor deck as a giant assembly table. This technique of prefabricating the trusses is easier and more accurate than other methods.

The roof of this 8-ft. by 12-ft. shed is framed with eight trusses spaced 16 in. o.c. The rafters and bottom chords are cut from 2×4s. Because this is a saltbox roof, the rafters along the rear of the roof are longer than the rafters at the front.

The three 2×4s that make up each truss are held together with plywood gusset plates that are glued and nailed across the joints on each side of the truss. Therefore, each truss requires six gussets: two at the peak, where the rafters meet, and two at each joint, where the ends of the bottom chord join the rafters.

CUT THE TRUSS PARTS

The quickest, most accurate way to assemble the trusses is to build them directly on the floor deck, using the square corner of the floor to align the rafters.

1. Start by cutting eight 2×4 rafters to 100 in. long and eight more to 55 in. Cut one end of each rafter at a 45-degree angle (known as a plumb cut) and cut the other end square.

2. Cut eight bottom chords to 64 in. long, mitering both ends of each board at a 45-degree angle.

3. Then saw all the gussets out of ½-in. ACX plywood. Cut each of the 16 triangular ridge gussets to 9 in. high by 18 in. wide; cut the 32 side gussets to 10 in. by 10 in.

B

C

D

INSTALL THE STOP BLOCKS

1. Cut seven 8-in.-long stop blocks from 1×3s.

2. Screw three of them vertically to the edge of the floor frame to support the 100-in.-long rafter, then move around the corner and fasten two more blocks to support the shorter rafter, as shown in **A**. (See the drawing on the facing page for the exact placement of the stop blocks.) Be sure to leave at least 2 in. of the 1×3 stop blocks protruding above the plywood floor deck. You're now ready to assemble the trusses.

ASSEMBLE THE TRUSS

1. On the floor deck, set a pair of rafters in place and press them up against the stop blocks, making sure the 45-degree miters fit together tightly at the peak, as shown in **B**. Then lay the bottom chord in place, making sure you have the proper overhang of 6¼ in. at the short rafter, and screw the remaining two stop blocks to the floor deck to hold the bottom chord in place, as shown in **C**.

2. With the seven stop blocks holding the three boards in position, glue and nail on the gusset plates, using construction adhesive and 1-in. roofing nails, as shown in **D**. Flip over the truss and install the gussets on the other side, then move the completed truss off to one side and repeat the process for the remaining seven trusses. When you're done, don't forget to remove the stop blocks.

4. Finally, install the remaining two intermediate blocks between the corner blocks at the front of the shed. Follow the same procedure, first removing the grass beneath the blocks and leveling them side to side, then use a 4-ft. level and a long 2×4 to level the blocks in the front row with the ones in the rear row, as shown in **D**.

Frame the floor

The structural frame of this shed's floor is made entirely from 2×6 and 2×8 pressure-treated lumber. Start by cutting two 2×8 mudsills and two 2×6 rim joists to 12 ft. long. Next, nail the 2×8s to the 2×6s with 3½-in. (16d) galvanized nails to form two L-shaped assemblies; space the nails about 12 in. apart. Set the assemblies on top of the foundation blocks running across the front and rear of the shed. Make sure the 2×8 mudsills rest on the blocks.

1. Next, cut ten 2×6 floor joists to 92⅞ in. long and set them between the two rim joists and on top of the mudsills, as shown in **E**. It's important that all the joists be cut to exactly the same length and that both ends of each joist be perfectly square.

2. Space the joists 16 in. o.c. and secure them with three 3½-in. (16d) galvanized nails, driving the nails through the front and rear rim joists and into the ends of the floor joists, as shown in **F**.

3. Next, set the two intermediate blocks between the two corner blocks at the rear of the shed, spacing them 42 in. o.c. Again, remove the grass beneath the intermediate blocks, compact the soil, and level each one, then set the long 2×4 and 4-ft. level across the four blocks, as shown in **C**.

After the joists are installed, check the floor frame for square by measuring the diagonals. If the two dimensions don't match, use a sledgehammer to lightly tap one corner of the longest dimension toward the center of the frame.

E

F

G

When the two diagonal dimensions are equal, the frame is square.

3. Secure the floor frame to the ground with four steel-cabled ground anchors, which are required by code in some areas Bore a $\frac{5}{8}$-in.-dia. hole through the mudsill near each corner of the frame, then insert the anchor's threaded rod through the hole and put on a washer and hex nut, as shown in **G**. Use a steel rod and hammer to drive the hold-down spike into the ground. If necessary, fine-tune the cable's tension by tightening the hex nut on the anchor's threaded rod.

Install the floor

To make the plywood floor, crosscut one sheet of $\frac{3}{4}$-in. tongue-and-groove ACX plywood in half to create two 4-ft. by 4-ft. pieces.

1. Starting in the front left corner of the floor frame, secure the half sheet to the floor joists with 2-in. decking screws spaced no more than 12 in. apart, as shown in **H**.

H

TRADE SECRET

When installing steel-cable ground anchors, it's important to drive the pointed hold-down spike deep into the ground. If the anchors didn't come with a drive pin, substitute a length of $\frac{1}{2}$-in.-dia. rebar. Insert the rebar into the spike, then use a sledgehammer to drive the spike into the ground.

PROTIP

Before nailing down the plywood floor, check the floor frame for square—one last time—by measuring the two corner-to-corner diagonals.

TRADE SECRET

Much of the shed's framing is screwed together using a cordless drill/driver and square-drive screws, not Phillips-head screws. Square-drive fasteners, also called Robertson screws, have a square hole, not a cross-shaped slot, milled into their heads; they're driven with a square-drive bit. The main advantage of square-drive screws is that it's virtually impossible for the bit to slip out.

2. Install a full plywood sheet next, making sure the tongue-and-groove joints fit together tightly. If necessary, use a sledgehammer and protective 2×4 block to tap the joint closed. Finally, screw down the two remaining pieces of plywood to complete the floor deck.

Wall Framing

When professional carpenters build houses, they typically build the walls on the flat floor deck, then tip them up and nail them in place. That same time-saving technique is employed here. Each of the four shed walls is framed, sheathed in plywood, and sided with cedar clapboards while it lies flat on the deck. Even the rake board and gable-end vents are installed ahead of time. When the front wall is set into place, the double-hung window will already be installed and the cedar trim will be nailed around the doorway.

Build a gable-end wall

Begin by building one of the two 8-ft.-wide gable-end walls. These end walls are framed entirely from 2×4s and have a saltbox profile, with the studs at the front of the walls longer than the ones at the rear. As a result, each wall has two top plates that slope up to the peak at a 45-degree angle. When building the end walls, keep in mind that you must also make a rough opening in the frame near the peak to install a louvered gable-end vent.

1. Cut the five 2×4s needed to form the perimeter of the wall, including the bottom plate, the end studs, and the two angled top plates. Screw the parts together with 3-in. decking screws, then begin cutting and installing the five intermediate studs, spacing them 16 in. o.c., as shown in **A**.

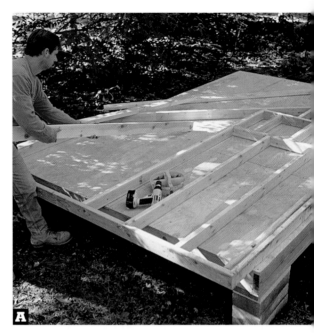

When you get to the center of the wall, measure 17¾ in. down from the peak to the top of the opening and frame in a 12-in. by 20-in. opening for the louvered vent. When the rough opening is complete, install the remaining wall studs.

After the wall is framed, cut two 2×4 rafters, making one 100 in. long and the other 55 in. long. These pieces are identical to the rafters you cut earlier for the trusses (see the sidebar on pp. 128–129). Set them on edge on top of the wall's top plates and screw them in place by driving 3-in. decking screws spaced 12 in. apart up through the plate and into the bottom edge of the rafter.

2. Hold the rafters flush with the outside edge of the plate. Be sure to fasten the longer rafter to the long rear top plate and the smaller rafter to the short front top plate, as shown in **B**.

Next, prepare to sheathe the wall frame with ½-in. ACX plywood. Lay a full 4-ft. by 8-ft. plywood sheet horizontally across the wall frame,

B

C

after—nailing in the studs. Cover the remaining wall frame with plywood, trimming it flush with the angled top plates.

Prepare the wall

Before you can begin nailing on the cedar bevel siding, you must first prep the wall. Cut off an 8-ft.-long section from a roll of 15-lb. felt underlayment (builder's paper) and slice the piece into 8-in.-wide strips.

1. Lay strips along both the left- and right-hand edges of the wall; these are the vertical edges that will form the corners of the building. Allow each strip to overhang the wall by 4 in., then attach it to the plywood sheathing with staples or 1-in. roofing nails, as shown in **D**. The asphalt-saturated paper will protect the plywood in case any water seeps behind the siding. Once the walls are erected, the paper's overhanging edge will be wrapped around the corner to protect the sheathing behind the corner boards.

sliding it down so it overhangs the wall's bottom plate by 7 in. Later, when you stand up the wall, the overhanging plywood will conceal the rim joist of the floor frame.

3. Align the ends of the plywood sheet flush with the wall frame, then fasten it with 8d nails spaced 10 in. to 12 in. apart, as shown in **C**. To square up the end wall before nailing off the sheathing, align the perimeter wall frame with the plywood deck before—and then again

D

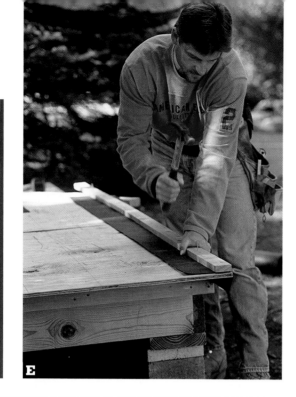

E

CORNER BOARD DETAIL

1/8-in. gap

Plywood wall sheathing

15-lb. felt underlayment

Bevel siding

Bevel siding

1×3 spacer strips

1×4 corner board

4¼-in.-wide corner board (ripped from a 1×6)

Hidden behind the 1×4 cedar corner boards is a pair of 1×3 spacer strips. The corner boards overlap the ends of the bevel siding and effectively cover the vertical seam between the siding courses and the spacer strips.

Next, cut four 8-in.-wide strips of felt underlayment to fit around the rough opening for the louvered vent and attach them with staples or 1-in. roofing nails. Run a continuous bead of silicone caulk along the back surface of the louvered vent's frame, set it in the rough opening, and attach it with six brass trim-head screws.

2. Before you install the siding, nail a long cedar or pressure-treated 1×3 to each edge of the wall, aligning it flush with the outer edge of the plywood sheathing, as shown in **E**. These 1×3s act as spacer strips and serve a very important function: After the walls are erected, the wider corner boards will overlap the 1×3s and conceal the vertical seam between the siding and the spacer strips.

Install the siding

Because this shed was going to be painted, I used 6-in. siding with a factory-applied coat of primer. If you're planning to stain your shed, use raw, unprimed clapboards instead.

1. Begin by cutting several pieces of siding to span the distance between the two 1×3 spacer strips. Don't be too concerned about cutting the boards to fit exactly in the space; the corner boards will cover up any small gaps. Next, make a starter strip by ripping a 2-in.-wide piece off the butt edge of one of the clapboards. The purpose of the starter strip is to bump out the first course of siding to match the angle of subsequent courses. Align the starter strip with the bottom of the plywood wall, but don't nail it in place yet.

2. Next, lay a piece of siding on top of the starter strip, allowing it to overhang the strip by ¼ in. to create a small drip edge, as shown in . Using 2½ in.-long ring-shank galvanized siding nails, nail through the siding and starter strip and into—but not yet through—the plywood sheathing. If you pound the nails home, the protruding points will prevent you from pressing the wall tightly against the floor frame when it comes time to raise the walls.

Install the second course of siding, overlapping the first course by 1 in. to leave 4½ in. of siding exposed to the weather. Again, tack-nail the siding to the plywood, being careful not to drive the nails too far.

3. When you reach the third course, you can begin driving the nails home. However, be sure to nail into a wall stud to avoid having dozens of sharp points sticking out inside the shed. Continue to cover the wall with siding, as shown in **G**.

When you come to the louvered vent, cut the siding to fit tightly against each side of the vent's frame. At the peak, trim the siding flush along the edges of the top plates.

Next, use siding nails to attach a cedar 1×4 rake filler strip to each of the rafters at the top of

the wall. Miter a 45-degree angle in the ends of the boards where they meet at the peak. Cut both 1×4s about 4 in. too long; the excess will be trimmed off after the fascia is installed.

4. Miter-cut a 45-degree angle in two cedar 1×6 rake boards and nail them on top of each filler board; allow the 1×6s to extend ¾ in. beyond the fillers, as shown in **H**.

PRO TIP

Each time you crosscut a piece of siding, brush a coat of stain or paint on the fresh end-grain cut to seal out moisture and deter rot.

WHAT CAN GO WRONG

If you're not paying close attention, it's very easy to install clapboards crookedly. Measure each course at both ends, or use a simple homemade gauge block, to ensure that the exposure to the weather remains consistent.

TRADE SECRET

To help position the walls and keep them straight, measure 3½ in. from all the edges of the floor deck and snap chalklines across the plywood before framing the walls. When it comes time to screw the wall to the floor, use the chalklines to help position the bottom plates. If there's a gap between the 2×4 plate and the line, have a helper push in the wall from the outside.

TRADE SECRET

When screwing the wall sections to the floor deck, position the 3-in. screws about 1 in. from the plywood sheathing. That way, the screws will go through the wall plate and plywood floor and into the 2×6 rim joist. Also, you'll find it much easier to drive the screws if you first bore ³⁄₁₆-in.-dia. screw-shank clearance holes.

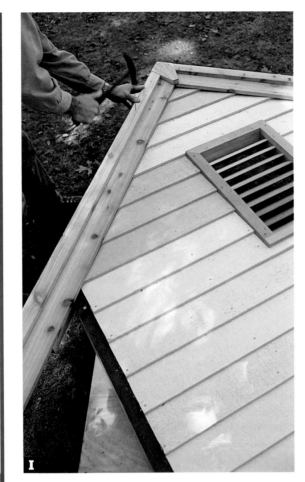

Make a decorative, emerald-shaped keystone block for the peak of the roof. Cut the 4-in.-wide by 5½-in.-high block from 5/4 cedar. Predill the block to prevent it from splitting, then fasten it to the rake boards with three 2½-in. ring-shank galvanized siding nails.

5. Next, cut two cedar 1×2s to length and set one on top of each 1×6 rake board. Align them flush with the top edges of the 1×6s, butt them tightly against the keystone, then nail them in place, as shown in **I**. These innocuous little boards, called rake trim, create additional shadow lines along the tops of the gable-end walls, which help emphasize the saltbox's roofline. Again, be sure to run the 1×6s and 1×2s a few inches longer than necessary.

Erect the first gable-end wall

Because these walls are heavy, you'll need at least one person to help you raise them. It's also a good idea to keep all kids and pets away from the site during this phase of the project.

1. Start by sliding the completed gable-end wall into position. Its bottom plate should be parallel with the end of the plywood floor deck and about 3 in. from the edge.

2. Grab the top of the wall by a 2×4 framing member—not the 1×4 rake board—and lift, as shown in **J**. When the wall is nearly vertical, have your helper slide over to the center of the wall while you move around to the other side. Push the very bottom of the wall sheathing tightly against the floor frame, then have your helper on the inside secure the plate to the floor deck with 3-in. decking screws.

For now, drive just one screw down through each end of the bottom plate. When the wall is pinned to the floor, hold it upright with two diagonal 2×4 braces. Finish screwing the bottom wall plate to the floor, starting in the middle of the wall. Have your assistant push in the center of the wall from the outside, then drive a 3-in. screw down through the wall plate. Drive two equally spaced screws between each pair of studs and down through the bottom plate.

3. Step around to the outside of the wall and hammer in the nails sticking out from the first two courses of siding, as shown in **K**.

Build the remaining walls

The next step is to build and erect the short rear wall, which is only 50¼ in. tall. Start by cutting the 2×4 top and bottom wall plates to 11 ft. 5 in. long. Cut ten 2×4 wall studs to 47¼ in. tall. Next, saw a 45-degree bevel along the top edge of the top wall plate. This is necessary to accommodate the rear rafters that come down from the peak at a 45-degree angle and rest on top of the rear wall.

Screw together the 2×4 wall frame with 3-in. decking screws; set the studs on 16-in. centers. Then install the plywood sheathing, spacer strips, starter strip, and bevel siding just as you did for the gable-end wall.

1. Remove the rear diagonal wall brace from the end wall, then lift the wall into position, as shown in **L**. Secure it by screwing down through the bottom plate. Use a 4-ft. level to check the corner where the rear wall joins the gable-end wall. Push the walls in or out until both are perfectly plumb, then fasten them together by driving five equally spaced 3-in. screws through the rear wall frame and into the end wall.

2. Now it's time to build another gable-end wall. The second one is identical to the first, except it's a mirror image. Use the same construction techniques as described previously. Don't forget to install a louvered vent and run the cedar rake boards and rake trim a little long.

The front wall is the same length as the rear wall—11 ft. 5 in.—but it's 82¼ in. tall. This wall has two rough openings: one for the door and one for the window. Start by cutting the 2×4 top and bottom wall plates to 11 ft. 5 in. long, then cut nine 2×4s to make 79¼-in.-tall wall studs. Begin screwing together the 2×4 wall frame with 3-in. decking screws; space the studs 16 in. o.c.

TRADE SECRET

After raising the gable-end wall and screwing it to the floor deck, you must install two diagonal 2×4 braces to prevent the wall from falling over. Make each temporary wall brace from an 8-ft.-long 2×4. Screw one end of the brace to the end wall stud about 5 ft. above the floor deck, then extend the brace diagonally and screw the other end to the rim joist. Add a second 2×4 brace on the opposite end of the wall.

The 34-in.-wide by 74-in.-high rough opening for the door is located 14¾ in. from the left end of the wall. Form the opening with two jack studs, a rough header, and four short cripple studs that fit between the header and the top plate.

The 25-in.-wide by 41½-in.-tall rough opening for the window is located 50 in. to the right of the door's rough opening. Just as you did for the door, install a rough header at the top of the window's opening. However, you'll also need to install a rough sill to establish the bottom of the opening. Cut a 5¼-in.-tall cripple stud to fit over the header and a 29½-in.-tall cripple to fit under the sill.

Once you've completed the frame, check it for square by measuring the diagonals, then nail the ½-in. ACX plywood sheathing to the framing with 2-in. (6d) nails. Cut 8-in.-wide strips of 15-lb. felt underlayment and nail them to the plywood around the door and window openings.

3. Then carefully set the double-hung window in the rough opening, as shown in **M**. Secure the window with 2 ½-in. (8d) galvanized nails.

4. Cut three pieces of roughsawn 1×4 cedar to fit around the door opening. Attach the trim boards, called casing, with 2½-in. siding nails, making sure they pass through the plywood sheathing and into the 2×4 wall frame.

5. Next, install the bevel siding just as you did for the other walls, maintaining a 4½-in. exposure. When you come to the door trim or window casing, cut the siding to fit tightly against each side. Lift the front wall into place, standing it between the two gable-end walls, as shown in **N**. Secure it by driving 3-in. screws down through the bottom plate and through each corner stud. After the wall is secured in place, use a handsaw to cut out the section of the bottom wall plate that runs across the doorway.

Roof Framing

If you took the time earlier to build the roof trusses before framing the walls, you're about to be rewarded. Framing the shed roof in the traditional manner—cutting and handling individual rafters, collar ties, and a ridge board—can take an entire day. However, it takes only about an hour to set all eight of the prebuilt trusses.

These trusses aren't very heavy, but you'll still need a helper to install them. Set up a stepladder inside the shed so one person can work along the top of the front wall. The second person must hand up the trusses to the first person, then go inside to screw the trusses in place.

Raise the trusses

The trusses are spaced 16 in. o.c., but there's no need to measure. All you need to do is set each truss directly over a wall stud. This greatly simplifies the installation process and, more important, helps transfer the weight of the roof to the wall frame and down to the sturdy floor frame and foundation blocks.

1. Hand the first truss to your helper on top of the wall, then hustle inside the shed and grab the long rear rafter of the first truss, as shown in **A**. Help your partner set the truss on top of both the front and rear walls, making sure it's

directly over the first wall stud. Working at the front wall, align the outside edge of the plywood gusset plate on the truss with the outside edge of the wall's top plate.

2. Drive two 3-in. screws up through the top plate and into the bottom of the truss. Because the stud is directly beneath the truss, you'll have to drive the screws up at a slight angle, placing one screw on each side of the stud, as shown in **B**. Don't fasten the truss to the rear wall just yet.

Continue installing trusses in this manner, making sure you place one over every stud. Attach the trusses to the front wall only and

PRO TIP

Outfit your cordless drill/driver with a handy drill-and-driver bit that allows you to switch quickly from boring pilot holes to driving screws.

PRO TIP

Conserve battery power when using a cordless drill/driver by rubbing paraffin onto the threads of the screws so they'll go in easier with less friction.

TRADE SECRET

**Aluminum soffit vents
have** either small slots
(little louvers) or tiny
holes that allow air to
flow through. For the
Saltbox Potting Shed, I
used 6-in.-wide vents. If
you can't find this size,
you'll have to improvise.
Go to a lumberyard and
order 12-in.-wide perfo-
rated aluminum soffit
vents. Then, use a utility
knife to score the panel
lengthwise along the
preformed bead that runs
down the center of the
vent. Carefully bend the
panel until it snaps into
two 6-in.-wide strips.

make sure the edge of each gusset plate lines
up with the top plate. After all eight trusses are
installed, begin fastening them to the rear wall.
However, this time start with the truss in the
middle of the roof.

3. On one of the gable-end walls, measure
how far the rear rafter tail overhangs the rear
wall, measuring from the end of the rafter to
the siding on the rear wall (it should be about
3 in.). Then move to the middle of the shed
and have your helper stand outside and measure
how far the rafter of the middle truss overhangs
the rear wall.

Push in or pull out the wall until the overhang
is equal to the one on the gable-end wall; this
will ensure that the rear wall is straight. Drive
two 3-in. screws up at an angle through the beveled top plate and into the bottom of the rafter.
With the middle truss securely holding the rear
wall, attach the rear rafters of the remaining seven
trusses, checking for a consistent overhang at
each rafter tail.

4. Next, fasten a long pine 1×4 across the rafters at the peak to hold the tops of the trusses
in position. Start by measuring the distance
between the gable-end walls and cut the 1×4 to
match. Hold the 1×4 against the rafters, close to
where they're screwed to the front wall. Mark
the center of each rafter on the 1×4. Climb up
to the peak and attach the 1×4 to the rafters
with 2-in. screws. Hold the top edge of the
pine board even with the peak, align each rafter
with its centerline, then drive in the screws, as
shown in **C**.

Add soffit vents

The rafter tails extend beyond the front and
rear walls by about 6 in., which is just enough
to form a small eave. The open space beneath
each eave is closed off with a standard 6-in.-
wide perforated aluminum soffit vent. The
vents allow fresh air to flow into the shed, then
exit from the wooden louvered vents mounted
on each gable-end wall.

D

E

F

1. To form each eave, start by cutting a long 2×4 to span the distance across all the rafter tails, including the ones on the gable-end walls. Nail the 2×4—called the subfascia—to the ends of the rafter tails with 3½-in. (16d) nails, as shown in **D**.

2. Slip the 6-in.-wide vent underneath the eave. Hold the vent at an angle, with its upper edge against the very top of the wall and its bottom edge flush with the 2×4 subfascia. Using 1-in. screws spaced 12 in. apart, attach the upper edge of the aluminum vent to the shed wall, as shown in **E**. Attach the lower edge of the vent to the bottom of the 2×4 with the same size screws.

3. Next, cut two long pine 1×4s to span the distance across the roof at the eave. Set one 1×4 on top of the rafters, align its lower edge with the outside face of the subfascia nailed across the rafter tails, then fasten it with two 2½-in. (8d) nails at each rafter. Set the second 1×4 on top of the rafters, butt it tightly against the first 1×4, then nail it in place. Use a handsaw to trim the 1×4 rake filler board flush with the outside surface of the subfascia at each end of the eave, being careful not to cut into the back of the 1×6 cedar rake board.

4. Cut a fascia board from a rough-sawn cedar 1×6 and nail it to the subfascia with 2½-in.-long ring-shank galvanized siding nails, as shown in **F**. Again, use a handsaw to trim the 1×6 rake boards flush with the fascia. Next, nail a 1×2 cedar fascia trim to the fascia. Cut the overhanging end of the 1×2 rake trim even with the fascia trim.

After completing the fascia trim on both roof edges, temporarily screw long 2×4s along the edges of the roof and flush with the top of the 1×2 fascia trim. These boards will act as alignment blocks for installing the starter course of cedar shingles.

Install spaced sheathing

The cedar roof shingles on this shed are installed in the traditional manner: over spaced sheathing, not plywood. This centuries-old building technique consists of a series of pine 1×4s nailed horizontally across the rafters. A 1½-in. space between the boards allows air to circulate behind the shingles. The width of the sheathing boards and the spaces between them are directly related to the amount of shingle exposed to the weather. In this case, the shingles have a 5-in. exposure.

1. Start by cutting nine 1×4s to span the distance across the roof. Rip one board to 2½ in. wide, set it on the rafters, and slide it tightly against the two 1×4s nailed on earlier. Fasten this board to each rafter with two 2½-in. (8d) nails, as shown in **G**. The reason you need to have three boards butt together along the eave is to provide adequate nailing support for the first three courses of shingles.

2. Next, set one of the 1×4 sheathing boards across the rafters. Hold it 1½ in. above the 2½-in.-wide board and nail it in place, again securing it to each rafter with two 2½-in. nails. Install the remaining 1×4s in a similar manner, working your way up the roof to the peak, as shown in **H**. Now move around to the back of the shed and install spaced sheathing on the rear rafters in the same way.

Roofing

You'll need approximately two squares (200 sq. ft.) of 16-in.-long cedar shingles for this saltbox roof. This includes a little extra to cover any waste. When ordering the shingles, be sure to ask for Number 1 Blue Label shingles, which are cut from all-heartwood timber for optimum decay-resistance.

Fasten each shingle with two—and only two—1½-in.-long galvanized shingle nails. These thin-shank nails are similar to siding nails, but they have a slightly larger head. Shingles come in random widths, so be sure to mix and match pieces of various widths as you work your way across the roof. Also, be sure to stagger all vertical seams by at least 1½ in. from one course to the next.

Nail down the shingles

The first course of shingles is commonly called the starter course, though I've heard some carpenters refer to it as the soldier course.

ROOF SHINGLE DETAIL

2×4 rafter

The cedar roof shingles are nailed over 1×4 spaced sheathing. Three boards are butted together at the eave. After that, the 1×4s are spaced 1½ in. apart. The first course of shingles is nailed 5 in. from the butt edge; all subsequent courses are nailed 6 in. from the butt edge.

2½-in. nails

1½-in. shingle nail

Cedar shingle

1×4 spaced sheathing

1½-in. space

1×4 spaced sheathing

1½-in. space

Cedar shingle

6 in.

2½-in.-wide pine board (ripped from a 1×4)

1×4

¾ in. (typical)

6 in.

5-in. exposure

Fourth course

1×6 fascia

1×4

6 in.

Third course

1×2 fascia trim

2×4 temporary alignment block

6 in.

5-in. exposure

5 in. (starter course only)

¼-in. overhang

¼-in. gap

Second course

1½-in. shingle nail

The starter course is flush with the alignment block.

Continue installing the starter-course shingles across the roof in this manner, as shown in **A**. When you reach the end, cut the last shingle to fit with a sharp utility knife, remembering to leave a 1-in. overhang.

2. The second course of shingles is installed directly on top of the starter course. This time, let the shingles hang down past the starter shingles by approximately ¼ in. That tiny overhang creates an effective drip edge. Measure 6 in. up from the butt edge and drive two nails per shingle. Install the rest of the second-course shingles, maintaining the ¼-in. overhang and nailing 6 in. from the butt.

3. Before installing a third course, measure 5 in. up from the butt edge of the second course and snap a line across the roof. Set the shingles for the third course on this line and nail them 6 in. up from the butt edge. Continue installing shingles in this manner, making sure you always keep a 5-in. exposure to the weather and locating the nails 6 in. up from the butt edge. This way, all the nail heads will be covered by the next course, as shown in **B**.

As you make your way across and up the roof to the peak, remember to stagger the vertical seams by at least 1½ in. When you reach the last course, you can either cut each shingle flush with the peak before nailing it down or let them all run long, then return later and trim them flush with a utility knife. At the rear of the shed, use the same techniques to shingle the long back slope of the roof.

Install a ridge cap

To finish the roof, build a continuous ridge cap from a cedar 1×4 and 1×6.

WHAT CAN GO WRONG

If you come across a cedar shingle that's 14 in. wide or wider, slice it in half with a utility knife. This is necessary because very wide shingles have a tendency to bow and curl. Also, it's customary to leave a ¼ -in. space between shingles. However, if the shingles are wet, space them ⅛ in. apart instead. When they dry out and shrink, you'll end up with ¼-in. spaces.

1. Starting at the lower left end of the roof, set the first shingle in place. Hold its butt (bottom) edge even with the 2×4 alignment block and allow its left edge to overhang the rake trim by 1 in. Measure 5 in. up from the butt edge and fasten the shingle with two nails, driving them 1 in. from the edge of the shingle. Set the next shingle beside the first one, leaving a ¼-in. space between them. Hold the shingle flush with the alignment block, then nail it 5 in. up from the lower edge.

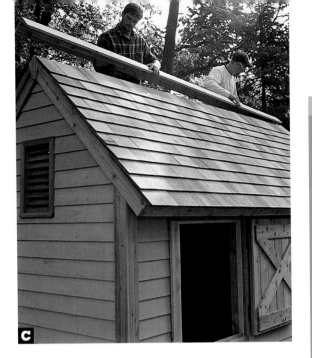

1. Measure the length of the upper course of shingles at the peak; it should be about 12 ft. 9 in. long. Add 1 in. to that dimension and crosscut the two boards to length.

2. Then use a portable circular saw or a table-saw to rip the 1×6 to 4¼ in. wide. Using 2½-in.-long siding nails, fasten the ripped 1×6 to the 1×4 to create an L-shaped cap that's 4¼ in. wide on each side.

3. Set the cap on top of the peak, making sure it overhangs the shingles by ½ in. at each end, as shown in **C**. Fasten the ridge cap to the roof with 2½-in.-long siding nails, driving two nails through the cap and into every other truss.

Dutch Door

Nearly everyone who has seen this saltbox shed has had the same initial reaction: "Wow, I love that door!" There's something about a Dutch door that captures people's attention and makes them smile. Of course, other types of doors can be used, but for this shed I built and installed a Dutch door. (For more information on building a Dutch door, see the sidebar on p. 146.)

Installing Corner Boards

AFTER THE ROOF HAS BEEN shingled, it's time to apply cedar trim to the four corners of the shed. As with the ridge cap, I find it easier to nail together two boards first to form a pre-assembled corner cap. Each corner board is made from a rough-sawn cedar 1×4 and 1×6. You must rip the 1×6 to 4¼ in. to create a "square" corner.

The top end of the ripped 1×6 must be mitered at a 45-degree angle to match the slope of the roof. To determine the length of this piece, measure from the butt edge of the first course of siding up the end wall to the rake board. For the 1×4, measure from the first course of siding up the front wall to the underside of the aluminum soffit vent, then cut the piece square at both ends. Nail the mitered board to the 1×4 with 2½-in.-long siding nails, spacing the nails 12 in. apart to create the L-shaped corner trim.

At each shed corner is a flap of felt underlayment that was installed before the siding was attached. Bend the flap around the wall corner, smooth any wrinkles, then fasten it with staples or roofing nails. Now slip the corner board into place, making sure it fits tightly against the soffit vent and rake board. Secure it with 2½-in.-long siding nails spaced 12 in. apart.

Fasten the preassembled corner boards to the front wall and the end walls with 2½-in.-long siding nails spaced 12 in. apart.

Building a Dutch Door

THIS DUTCH DOOR IS BASICALLY A batten door, so its construction is straightforward. The entire door is made from V-jointed, tongue-and-groove cedar 1×6 boards. Some of the boards are ripped to 4 in. wide and used to make the vertical stiles, horizontal rails, and X-shaped cross bucks. The door is initially constructed as a single unit, measuring 33½ in. wide by 74 in. high. Then it's cut into two panels, with the bottom half 6 in. taller than the top half.

1. Cut seven tongue-and-groove 1×6s to 74 in. long. Then rip the tongue off one board and the groove off another. Lay the boards on a workbench, with their rough-sawn surfaces facing down, as shown in **A**, and join the boards together to form a panel, making sure you start and end with a square-edged board.

2. Cut four horizontal rails from one of the 4-in.-wide cedar boards. Place two of the rails 2 in. from each end of the door, and place the other two side by side across the middle of the door. Adjust them until the seam between these two center rails is 40 in. from the bottom. That's where you'll cut the door into two panels. Measure and cut the 4-in.-wide pieces to create the cross-buck design, then attach them with construction adhesive and 1¼-in. decking screws.

3. Now it's time to flip over the door and use a circular saw to crosscut it into two panels, using a straightedge clamped to the door as a saw guide. Build a perimeter face frame for each door panel out of 4-in.-wide cedar, setting the vertical stiles between the horizontal rails and attaching each board with construction adhesive and 1¼-in. (3d) nails. Cut and fit the X-shaped cross bucks, as shown in **B**. Then install them using the same adhesive and 1¼-in.-long nails, as shown in **C**.

A

B

C

Install the door

Because a Dutch door has two panels, the installation is slightly more complicated than for that of a standard door. To help simplify the task, I used surface-mounted hinges and enlisted the help of a friend. You'll need two 4-in. hinges for each door panel; the hinges are screwed to the door panels before the door is installed. I like to position the hinges 4 in. from the top and bottom edges of the panels.

1. Starting with the upper panel, hold it in place at the top of the door opening and press it tightly against the right-hand side casing. To ensure that there's proper clearance at the top of the door, slip a ¼-in.-thick shim between it and the 1×4 cedar head casing nailed over the doorway. Bore ³⁄₃₂-in.-dia. pilot holes through the hinge-leaf holes and into the side casing, then attach the upper panel with the weather-resistant screws that come with the hinges, as shown in **A**.

2. Install the lower door panel in the same way. First, hold it in place at the bottom of the door opening, using ¼-in.-thick shims between the two door panels to create sufficient clearance. Slide the panel tightly against the right-hand side casing, then attach each hinge leaf with four screws, as shown in **B**.

On the inside of the door's lower half, install a 4-in. sliding barrel bolt. Place the bolt flush with the top edge of the door and about 5 in. from the latch edge. Mount the mating part of the bolt on the inside of the door's upper half.

A

B

PRO TIP

To operate a Dutch door as a single unit, install a sliding barrel bolt vertically on the inside surfaces to hold together the two door panels.

PRO TIP

This door's cross-buck face frame is fastened with 1¼-in. nails. If you don't have this size nail, use wire cutters to snip longer ones down to size.

WHAT CAN GO WRONG

The boards you use to build the doors must be thoroughly dry. If they're even the slightest bit damp, the doors will shrink as the wood dries, resulting in an ill-fitting door. Stack the boards indoors with "stickers" (wood strips) in between to allow air to circulate around each board. Wait at least a week before building the doors.

A Dutch door is built as a single unit, then cut in two. However, this cut is too critical to make freehand, so you must use a straightedge guide. Cut the guide from a rip of plywood with a factory edge, or use a straight 1×6 board. Clamp the guide to the door, making sure it's perfectly parallel with the cut line. The distance from the guide to the cut line must equal the distance from the sawblade to the edge of the saw's shoe.

Simplify the job of hanging the two door panels by using surface-mounted hinges, which don't require you to cut precise mortises for each hinge leaf.

Entry Deck

The plywood floor of this shed ended up being about 20 in. above the ground, just high enough to accommodate a two-tier wooden platform step. I could have made a set of wooden stairs or even a ramp, but a platform step is more attractive and easier to build. If the shed floor had been about 14 in. above the ground, I would have needed only a single platform.

Each platform consists of a 2×6 pressure-treated frame topped with 5/4 cedar boards, which measure $1\frac{1}{16}$ in. thick and $3\frac{1}{2}$ in. wide. If your local lumberyard doesn't carry 5/4 boards, use cedar or pressure-treated 2×4s instead. Build the bottom platform at least 16 in. longer than the width of the doorway opening. Make the top platform equal in length to the width of the doorway and 10 in. shallower than the bottom platform to create a wide, comfortable step.

Set the concrete blocks

To prevent the steps from sinking into the ground, set the bottom platform on top of four 2-in.-thick concrete patio blocks.

1. Smooth the dirt in front of the doorway, then arrange the blocks so that one supports each corner of the platform.

2. Place a 4-ft. level across the blocks and check for level in two directions: front to back and side to side, as shown in **A**. If necessary, stack two patio blocks on top of each other to level them.

Build the platforms

The bottom platform measures $27\frac{1}{4}$ in. deep by $46\frac{1}{2}$ in. wide.

1. Start by cutting two 2×6s to $46\frac{1}{2}$ in. long; these pieces will be the front and back of the frame. Cut four 2×6s to $24\frac{1}{4}$ in. long to use as the left and right ends and the two intermediate supports.

2. Screw the front and back pieces to the two ends with 3-in. decking screws. Slide the two support boards between the front and back, space them 16 in. o.c., then screw them in place.

3. Set the frame on top of the concrete blocks and check to make sure it extends past the doorway an equal amount on each side. Cut eight 5/4 by 4-in. boards to 48 in. long and nail the boards to the platform with $2\frac{1}{2}$-in. siding nails, as shown in **B**. Allow each board to overhang the ends of the frame by $\frac{3}{4}$ in.

4. Next, build the 2×6 frame for the top platform, which measures $16\frac{3}{4}$ in. deep by $30\frac{1}{2}$ in. wide. Cut two 2×6s to $30\frac{1}{2}$ in. long for the front and back of the frame and two to

13¾ in. long for the left and right ends, then screw the frame together with 3-in. screws. Now cut five 5/4 by 4-in. boards to 32 in. long and nail them to the platform. Set the top platform on the bottom one and slide the assembly against the shed, as shown in **C**.

5. Finally, fasten together the two platforms by driving two 3-in. screws down at an angle through each side of the top platform and into the bottom platform, as shown in **D**. Also, screw the top platform to the shed's floor frame to prevent the steps from shifting out of position.

Potting Bench

No potting shed is complete without a potting bench, which serves as a gardener's workbench. It provides a convenient place to mix soil, trim plants, and fill pots. You can buy fancy potting benches at most garden shops, but I decided to make one from ¾-in. ACX plywood. The bench is supported 36 in. off the floor by four wooden brackets, which are securely screwed to the wall studs for maximum strength. The bench measures 24 in. wide by 91¾-in. long, but you can make a longer one if you want; just be sure to install more brackets.

Build the brackets

Each bracket consists of two horizontal 1×4s and one diagonal 2×4 brace. Cut the 1×4s to 25½ in. long, mitering one end of each piece at a 45-degree angle. Cut each 2×4 brace to 31½ in. long, mitering both ends at a 45-degree angle.

WHAT CAN GO WRONG

A potting bench is installed along the shed's rear wall, directly below the low-slung saltbox roof. If you make the bench too narrow, you won't be able to stand at it without hitting your head against the 2×4 rafters. Cut the plywood for the bench 24 in. deep, which will force you to stand far enough back from the wall to safely clear the rafters.

PRO TIP

When you buy the plywood for the potting bench, have the salesperson cut it to size. Most lumberyards and large home centers will cut it for free.

POTTING BENCH BRACKET

The plywood potting bench is supported by wooden brackets, which are screwed to the wall studs. Each bracket is made of two horizontal 1×4s and a diagonal 2×4 brace.

25½-in.-long 1×4 (two required per bracket)

½ in.

1⅝-in. screws

31½-in.-long 2×4 brace

2½-in. screw

35¼ in. above floor

45 degrees

2×4 wall stud

TRADE SECRET

Protect the plywood potting bench from water and soil stains by applying three coats of exterior-grade polyurethane varnish. Allow each coat to dry overnight, then lightly sand the surface with 120-grit sandpaper.

1. Using carpenter's glue and 1⅝-in. drywall screws, fasten the 1×4s to the 2×4, as shown in **A**.

2. Use a 4-ft. level to mark a level line across the rear wall studs 35¼ in. above the shed floor. Slip one of the brackets over the stud, making sure the top edges of the 1×4s are even with the pencil mark, as shown in **B**. Attach the bracket by driving three 1⅝-in. screws through each 1×4 and into the side of the stud. Also, drive one 2½-in. screw through the bottom end of the 2×4 brace and into the edge of the stud. Attach the remaining three braces in a similar manner, spacing them equally across the back wall.

Install the bench

The ¾-in. plywood used for the bench top is plenty sturdy enough to handle everyday potting chores, but be sure to use an exterior-grade plywood, which will resist warping and delaminating. For added protection, apply three coats of polyurethane varnish.

1. Cut the plywood bench to size with a circular saw and set it on top of the brackets. Then fasten the bench to each bracket by driving two 1⅝-in. screws down through the plywood and into the 1×4s below.

2. To give the bench a more finished look, you can cover the edges with a piece of ¾-in.-wide screen mold, which you can buy at any home center or lumberyard. Apply glue to the front edge and the end of the bench, then nail on the ¼-in.-thick screen mold with ¾-in. brads, as shown in **C**.

3. Complete the potting bench by installing a backsplash along the rear wall. This useful accessory will keep items from falling behind the back of the bench. To make the backsplash, cut a pine 1×4 as long as the bench and secure it to each stud with two 1⅝-in. screws, as shown in **D**.

Storage Idea

ON THE INSIDE OF THE front wall, between the window and the door, is a large blank section. Rather than see this perfectly good space go to waste, I converted it into a tool-storage area by installing a perforated hardboard panel (also commonly called pegboard). The panel provides a convenient place to hang various lawn and garden tools. At the top of the perforated panel, I installed two long brackets and a narrow strip of wood to create a shelf for storing clay pots.

The ¼-in.-thick panel measures 43½ in. wide by 54½ in. high; it's attached to the 2×4 wall framing with 1¼-in. drywall screws. Hold the panel against the wall studs, making sure it's tight against the underside of the trusses. Drive the screws into the wall's top plate at the top of the panel and into the studs along the panel's bottom edge and two side edges.

The panel is mounted right beside the doorway and offers more than 16 sq. ft. of space for hanging tools and supplies.

DESIGN
OPTIONS
BEYOND BASIC STORAGE

▲ ▶ **This 10-ft. by 12-ft. antique tool shed (above) was rescued from demolition and transformed into a private oasis, complete with whirlpool spa and red-cedar wall paneling (right).**

◀ **Built as a poolside studio, this charming 15-ft. by 20-ft. outbuilding features an arched door, cedar-shingled walls and roof, and double-casement windows crowned with half-round fanlight windows.**

This stately pool pavil-
ion houses changing
quarters, a bathroom,
and an interior suite
for relaxing. Note that
the roof extends about
6 ft. beyond the 12-ft.
by 14-ft. main struc-
ture to create a shady
retreat. Two round fiber-
glass columns support
the roof extension.

This 12-ft. by 16-ft.
lakeside cabin features
cedar vertical-board
siding, architectural
roof shingles, and a
small sundeck. Its large
windows and sliding-
glass doors flood the
interior with natural
light and provide great
views of the lake.

Measuring 10 ft. by
20 ft., this versatile
backyard building is
divided in half by a
partition wall. On the
right side is a cool and
cozy screen room. To
the left, behind the
arched door, is a kid's
playhouse, which could
also be used for storage.

COLONIAL-STYLE SHED

Aquintessential storage building, this 10-ft. by 16-ft. colonial-style shed provides more than enough storage space for the average household, yet it is compact enough to fit in the smallest backyard. Because the structure is less than 200 sq. ft., most building departments will allow you to build it on an on-grade foundation. This particular shed was set on a skid foundation formed with 6×6 timbers.

This shed features vertical-board cedar siding, large 2-ft. by 4-ft. barn-sash windows, and a pair of double-wide batten doors. The 10-in-12 roof slope is covered with architectural-style asphalt shingles. Note that the roof extends beyond each end wall by about 8 in. to create a gable overhang. This classic architectural detail, which is seldom found on storage sheds, emphasizes the gable roof and creates deep shadow lines at the end walls. (To order a set of building plans for the Colonial-Style Garden Shed, see Resources on p. 214.) ▶ ▶ ▶

Skid Foundation

This shed features a skid foundation that supports the building on two long, straight timbers laid on the ground in parallel rows. The timbers (skids) are leveled and then the shed's floor frame is fastened to them. For this shed, the skids were made of pressure-treated 6×6s, but 4×6s or 8×8s could have been used as well. It's important to make the skids from pressure-treated lumber rated for either ground-contact use (chemical retention level of .40 lb./cu. ft.) or wood foundations (.60 lb./cu. ft.).

Although the skids can be set directly on the ground, it's best to lay them on a bed of gravel for increased stability and long-term support. In most instances, you need to put only a strip of gravel underneath each skid, but in this particular case, gravel was spread over the entire building site because the area attracts a fair amount of groundwater runoff.

Set the skids

If you're building the shed on a relatively level lot, you'll need only two 6×6 skids; cut each one to 15 ft. 8 in. long. However, if the site slopes 4 in. or so—as this one did—you'll have to stack two 6×6s to build up the low end of the foundation.

SKID FOUNDATION

When the ground slopes, stack 6×6s to create a level skid foundation. The first tier is partially buried and secured with rebar pins driven into the ground. Full-length skids are set on top and screwed to the bottom-tier skids. For this particular shed, space the skids 7 ft. 4 in. apart.

6×6 skid, 15 ft. 8 in. long (top tier)

¾-in. dia. × 1-in.-deep counterbored hole

6-in. screw

24 in.

12 in.

7 ft. 4 in.

6×6 skid, 10 ft. long (bottom tier)

½-in.-dia. × 18-in.-long rebar pin

30 in. (typ.)

12 in.   6-in. screw   12 in.

Rebar pin

2-in.- to 4-in.-thick gravel bed

Ridge vent

Soffit vent

Plywood roof sheathing

2×4 roof rafter

Architectural-style roof shingles

Roof truss

Cedar keystone

Rake trim

Rake board

8-in. gable overhang

Cedar corner board

Batten door

Transom window

Plywood floor

6×6 skid

Skid foundation

Pressure-treated floor frame

2×6 floor joist

2×4 wall blocking

2×4 girt

Fascia board

Fascia trim

Barn-sash window

V-jointed tongue-and-groove siding

PRO TIP

Be sure to wear gloves,
eye goggles, and a
dust mask when sawing
treated wood to protect
you from the chemical-
laced dust.

WHAT CAN GO WRONG

Here's a common prob-
lem that, fortunately, has
a quick fix. Sometimes
the foundation's skids are
level, but when the floor
frame is installed, it's
slightly out of level. Don't
bother messing with the
skids; just slip a couple of
thin shims under the low
end of the frame.

PRO TIP

Don't assume that the
factory-cut ends of fram-
ing lumber are square.
Before framing any floors
or walls, check each one
with a framing square.

1. To make the stacking 6×6s, first cut two 6×6s to 10 ft. long for the first (lower) tier of skids, set them in position on the gravel bed, and check them for level. The distance across the skids, from outside edge to outside edge, must be 7 ft. 4 in.

2. Secure the skids to the ground by first boring three equally spaced ½-in.-dia. holes down through each 6×6. Then use a sledgehammer to pound a ½-in.-dia. by 18-in.-long rebar through each hole and deep into the ground.

3. Next, cut two 6×6s to 15 ft. 8 in. long and set them on top of the lower-tier skids so the ends and sides of the top-tier timbers are flush with those on the lower tier, as shown in **A**.

4. Then measure about 1 ft. in from the ends of the upper skids, bore 1-in.-deep holes with a ¾-in.-dia. spade bit, and drive a 6-in.-long screw through each hole to fasten the top-tier 6×6s to the lower ones, as shown in **B**. Drive four more screws, spaced 24 in. apart, through each upper-tier skid.

Flooring

The understructure frame of the shed's floor is made entirely of pressure-treated 2×6s. It consists of 13 floor joists and two rim joists, which span across the front and rear of the floor frame. Installed on top of the framing is a smooth and strong ¾-in. ACX plywood floor.

Assemble the frame

1. Cut the two 2×6 pressure-treated rim joists to 15 ft. 8 in. long, then cut the 13 floor joists to 9 ft. 9 in. long.

2. Start assembling the rectangular floor frame by first fastening the rim joists to two of the floor joists, driving three 3-in. decking screws at each corner joint.

3. Next, set the assembled perimeter floor frame on top of the 6×6 skids, with the long rim joists parallel to the skids. The 10-ft.-wide frame is 32 in. wider than the distance between the two skids, so make sure it overhangs each skid evenly by 16 in.

4. Hold the sides of the frame even with the ends of the skids, then drive two 3-in. screws at an angle down through a joist and into one of the skids, as shown in . With one corner screwed fast, check the frame for square by measuring across the two diagonals. If necessary, tap one corner with a sledgehammer to square up the frame (it's square when the measurements are equal). Finish attaching the frame to the skids with two screws at each of the remaining three corners.

5. Now set the remaining 11 floor joists between the rim joists and on top of the skids, as shown in **B**. Install the first 2×6 joist 12 in. o.c. from the end of the floor frame. Drive three 16d nails through the front and rear rim joists and into each end of the joist. Then install the remaining 10 joists, spacing them 16 in. o.c.

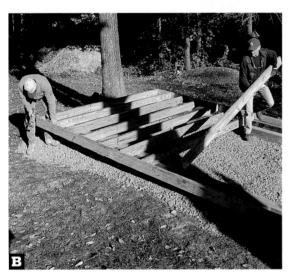

6. Check the frame for square one last time before fastening each joist to each skid with 3-in. screws. Drive the screws at an angle down through the joists and into the skids, as shown in **C**. If you find that the screws are splitting the joists, bore $\frac{3}{16}$-in.-dia. screw-shank clearance holes.

ACCORDING TO CODE

In some towns, the local building code will require you to secure the floor frame to the skids with more than just screws. In those cases, use metal tie-down straps (also called hurricane clips). Screw one end of the strap to the floor joist and the other end to the skid. One strap on each end of every other joist is usually sufficient.

Install the plywood floor

1. Install a full 4-ft. by 8-ft. sheet of ¾-in. ACX tongue-and-groove plywood along the rear edge of the floor frame. Align the sheet perpendicular to the floor joists, with its grooved edge flush with the rear rim joist. Fasten the sheet to the floor joists with 8d galvanized nails spaced about 12 in. apart. Then crosscut the next sheet to 92 in. long, set it end to end with the first sheet, and nail it down. Make sure the end seam falls on the center of a joist.

2. Crosscut a full-size sheet in half to form two 4-ft. by 4-ft. pieces. Cut one of those pieces to 44 in. and lay it edge to edge with the installed 92-in. sheet. If you have trouble closing the tongue-and-groove joint between the two sheets, gently tap them together with a sledgehammer and a protective 2×4 block, as shown in **D**. Nail the 44-in. piece to the joists, then install a full sheet, followed by the 4-ft. by 4-ft. half sheet. These three pieces add up to 15 ft. 8 in., the exact length of the floor frame.

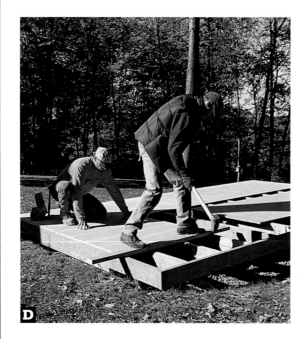

D

3. Rip the last full sheet of plywood lengthwise down the center to create two 2-ft. by 8-ft. strips. Crosscut one strip to 92 in. Finish covering the floor by nailing the two plywood strips to the joists along the front edge of the floor frame.

Before building and erecting each of the shed's walls, I use the large, clean plywood floor surface to lay out and build the seven roof trusses. (For more detailed instructions, see the sidebar on pp. 162–163.) It's much quicker and easier to build the walls on the floor deck, then tip them up into place.

Gable-End Trusses

The two gable-end trusses are very similar to the standard roof trusses (see p. 162). As with the other roof trusses, each one is assembled with two 87-in.-long rafters and a 10-ft.-long bottom chord. However, the gable-end trusses have two additional 2×4 structural supports, which are attached with plywood gusset plates.

After fastening siding to the trusses, build the rake overhangs before setting the assembly in place on the end walls. This entire process is done on the floor, making it a lot easier and safer than working on a ladder.

Install structural supports

The first structural support is a horizontal collar tie installed between the rafters 19 in. above the bottom chord. This 76¾-in.-long board adds rigidity to the truss and provides nailing support for the V-jointed cedar siding (see the photo on the facing page).

The second structural support is an 8-ft.-long shoe plate fastened flush with the bottom chord. When the truss is installed, the shoe plate rests on top of—and is screwed to—the top plate of the gable-end wall. When fastening the structural supports to the gable-end trusses, be sure to nail the plywood gusset plates on one side only—the side that faces in toward the center of the shed.

Form the gable overhang

Assemble the trusses (as described in the sidebar on pp. 162–163), then you'll be ready to make the gable overhangs.

1. Cover their exterior surfaces with V-jointed, tongue-and-groove cedar 1×6s, as shown in **A**. For now, let the boards run long at the top of the truss. Cut the siding to overhang the bottom chord by 1 in. Nail the siding to the rafters, bottom chord, and collar tie with 1½-in. ring-shank galvanized siding nails, driving the nails at an angle down through the tongue of each board. Using a circular saw, trim the cedar boards flush with the top of the rafters.

2. Next, build the gable overhang on each of the gable-end trusses. This overhang, also called a rake overhang, is essentially an 8-in.-deep eave framed with 2×4s. It's attached to the

GABLE OVERHANG

Roof shingles

1×2 cedar rake trim

5-in.-wide cedar rake board

2×4 blocking

Truss rafter

V-jointed, tongue-and-groove cedar soffit

Plywood roof sheathing

Cedar siding

8 in.

The roof extends past the front and rear walls of the shed by 8 in., creating a gable overhang. The 2×4 frame of the overhang is screwed to the gable-end truss. Its underside is covered with short pieces of cedar siding to form a soffit.

top of the truss, then its underside is covered with short pieces of 1×6 cedar siding to form a soffit.

Start by cutting four 87-in.-long rafters from 2×4 stock. Make a 40-degree miter on one end of each rafter. Cut ten short 2×4 blocks to 5 in. long. Stand five of the short blocks between two of the rafters, spacing them equally, then screw them together with 2½-in. decking screws to form a ladder. Assemble the remaining two rafters and five blocks to build the second frame. Attach the frames to the top of the truss with 3-in. decking screws. Drive the screws down through the frame and into the rafter.

3. Cut 8-in.-long pieces of V-jointed, tongue-and-groove cedar 1×6s—enough to cover the

A

Building Roof Trusses

40-degree plumb cut

½-in. plywood stop block

2×4 rafter

½ in.

87 in.

5½ in.

18 in.

2×4 bottom chord (ceiling joist)

36 in.

36 in.

120 in.

½-in. plywood
stop block

½-in. plywood
stop block

½-in. plywood side gusset
(both sides) (11 in. × 11 in.)

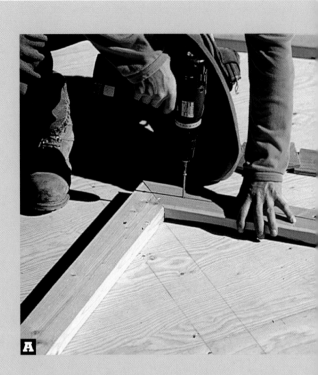

A

Roof-Truss Assembly

Each of the seven roof trusses is built from two 2×4 rafters and a 2×4 bottom chord. One truss is screwed to the plywood floor deck and used as a template for assembling the remaining trusses. Eight plywood stop blocks are screwed to the template truss; they hold the rafters and bottom chord in proper alignment. Plywood gusset plates are glued and nailed across the joints to hold each truss together.

Each roof truss is made from two angled 2×4 rafters and one horizontal 2×4 bottom chord (ceiling joist). The three parts are held together with ½-in.-thick plywood gusset plates glued and nailed across the joints. This particular roof is framed with seven standard trusses, which are spaced 24 in. o.c., and two gable-end trusses. These last two trusses are installed atop the end walls of the shed and, just like the walls, are sided with V-jointed, tongue-and-groove cedar 1×6s.

It's best to assemble the trusses right on the floor deck. However, because of this shed's roof pitch, you can't use the square 90-degree corner of the floor to align the rafters as you did for the Saltbox Potting Shed's roof trusses (see Chapter 6). That's because the rafters for this shed are cut at a 40-degree angle; when assembled, they form an 80-degree angle. But there's an alternative assembly method that's just as easy and accurate.

1. Start by cutting the truss parts to size. For each truss, cut two 2×4 rafters to 87 in. long. Cut a 40-degree miter on one end of each rafter; cut the other end square. Next, cut a 2×4 bottom chord to 120 in. long, mitering both ends at a 50-degree angle. Then saw the gusset plates from ½-in. ACX plywood. Cut each of the triangular ridge gussets to 6¾ in. high by 17 in. wide. Cut the side gussets to 11 in. square, then lop off the corner, as shown in the drawing above. Also, cut eight 4-in. by 4-in. plywood stop blocks.

2. In the center of the floor deck, lay out two rafters and a bottom chord to form one roof truss, temporarily screwing the parts to the floor with 3-in. screws, as shown in **A**. After making sure the joints between the boards fit tightly together, use 1⅝-in. screws to fasten the plywood stop blocks to the outside edges of the bottom chord and rafters, as shown in **B**. You'll also need to attach a stop block to the end of each rafter tail, as shown in **C**.

3. Now use this setup as a template for assembling the seven standard roof trusses. Set one pair of rafters on top of the rafters screwed to the floor deck, as shown in **D**. Press them against the stop blocks and make sure the 40-degree miters fit together tightly at the peak. Lay a bottom chord on top of the one screwed to the floor. With the eight stop blocks holding everything in position, glue and nail on the gusset plates, using construction adhesive and 1-in. roofing nails, as shown in **E**. Flip over the truss and attach gussets to the other side. Remove the completed truss and repeat the process for the remaining trusses. When you're done, remove the stop blocks and fasten gusset plates to the truss screwed to the floor deck. Then unscrew the truss, flip it over, and install gussets on the other side.

TRADE SECRET

Here's an easy way to increase the cutting capacity of a power miter saw. Lay a scrap 2×6 across the saw table, then place the board you're cutting on top. That raises the board to the widest part of the blade, which increases the cutting capacity of a 10-in. miter saw from about 5½ in. to 6½ in.

TRADE SECRET

After assembling the roof trusses, set them off to one side on top of a few scrap 2×4s; don't lay them right on the ground. Cover the stack with a waterproof tarp. If the trusses get wet, they'll be much heavier to lift into place.

B

entire underside of the overhang—and fasten them to the underside of the overhang with 1½-in. siding nails, as shown in **B**.

4. Cover the face of the overhang—and the ends of the soffit boards—with a 5-in.-wide rake board ripped from a cedar 1×6. Position the rake so it projects ½ in. above the top edge of the 2×4 overhang; the raised lip will hide the edge of the plywood roof sheathing. Next, make a decorative emerald-shaped keystone block for the peak of the roof. Cut the 4-in.-wide by 5½-in.-tall block from 5/4 cedar. Predrill the block to prevent it from splitting, and fasten it to the rake boards with three 2½-in. ring-shank galvanized siding nails. Nail a 1×2 rake trim piece to the rake board, keeping it flush with the top edge of the rake board and butted tightly against the keystone. Cut both the rake board and the trim piece about 6 in. too long, then trim them to size after the fascia is installed.

Walls

The walls of this shed are built on the floor deck, then tipped up and nailed into place, a time-saving technique practiced by many professional homebuilders. Each wall is framed with 2×4s and sided with V-jointed cedar boards while it lies flat on the deck. The window and door openings are framed in the walls, so there's no additional framing necessary after the walls are erected.

Start by building the left-hand sidewall out of 2×4s. This wall has the two 2-ft. by 4-ft. barn sash windows; the opposite sidewall has no windows. Note in the drawing on the facing page that the wall is framed with both vertical blocking and horizontal girts (or purlins). The pieces of blocking serve as wall studs to support the weight of the roof. They're spaced 24 in. o.c. and are cut to fit between the long horizontal girts, which span the entire length of the wall. The girts provide solid support for nailing up the vertical cedar siding; they also help keep the wall straight. Therefore, it's imperative that you use only perfectly straight 2×4s for the girts.

Frame the sidewall

1. Begin by cutting the 2×4 top and bottom wall plates to 15 ft. 8 in. long. Cut two end studs to 79¼ in. long, then fasten the wall plates to the end studs with 16d nails to create a rectangular wall frame. Rip one 3½-in.-wide strip of ½-in. plywood to fit against the inside surface of each end stud; attach the strips with 1¼-in. screws. After the wall is complete, these plywood spacer strips help create a sturdy 3½-in.-wide corner post at each end of the wall. Check the wall frame for square by measuring the opposing diagonals (when the two dimensions are equal, the frame is square).

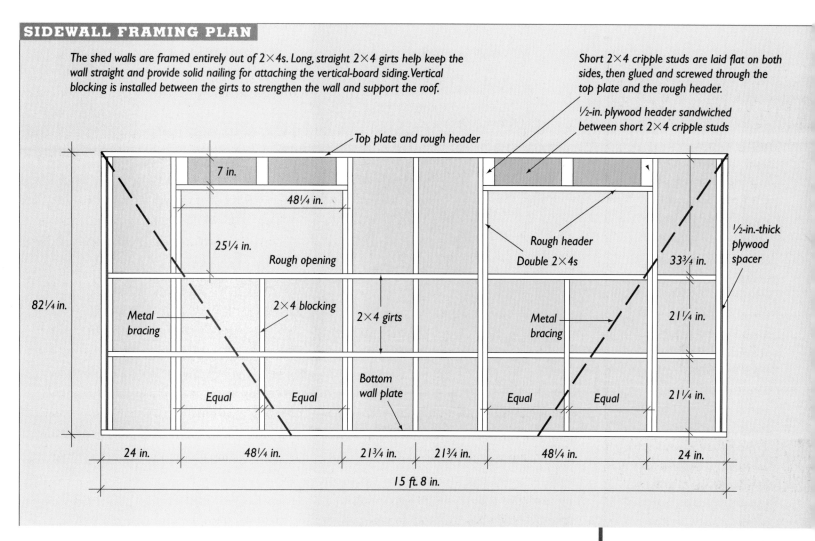

The shed walls are framed entirely out of 2×4s. Long, straight 2×4 girts help keep the wall straight and provide solid nailing for attaching the vertical-board siding. Vertical blocking is installed between the girts to strengthen the wall and support the roof.

Short 2×4 cripple studs are laid flat on both sides, then glued and screwed through the top plate and the rough header.

½-in. plywood header sandwiched between short 2×4 cripple studs

Top plate and rough header

7 in.

48¼ in.

25¼ in.

Rough opening

Rough header
Double 2×4s

½-in.-thick plywood spacer

33¾ in.

82¼ in.

Metal bracing

2×4 blocking

2×4 girts

Metal bracing

21¼ in.

Bottom wall plate

Equal   Equal

Equal   Equal

21¼ in.

24 in.   48¼ in.   21¾ in.   21¾ in.   48¼ in.   24 in.

15 ft. 8 in.

2. Next, cut 18 pieces of 2×4 blocking to 21¼ in. long. Stand nine of the pieces along the bottom wall plate, spacing them 24 in. o.c., and fasten each one with two 16d nails driven up through the bottom plate. Cut two 2×4 girts to 15 ft. 4 in. long (remember to use lumber that is as straight as possible).

3. Set one of the girts between the end studs and on top of the nine pieces of blocking fastened to the bottom plate, and fasten the girt to each 2×4 block with two 16d nails, as shown in **A**. (See the drawing above.)

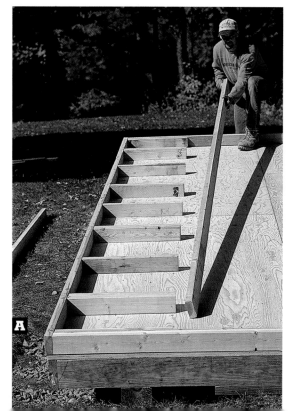

A

PROTIP

Use a 20-oz. or heavier framing hammer to build the walls. You'll be able to drive larger nails with fewer blows than you would with a 16-oz. hammer.

When building a wall with vertical blocking and vertical-board siding, you must install diagonal bracing to strengthen the wall frame. The simplest type to install is metal let-in bracing, sometimes called wind bracing. Snap two diagonal chalklines from the center of the bottom plate to a point on each end stud about 4 in. down from the top wall plate. Then make two diagonal cuts (about 1 in. deep) along the lines. Tap the metal bracing into the kerf and secure it with roofing nails.

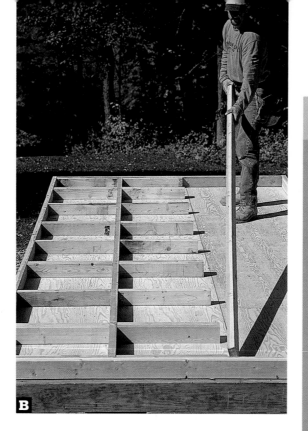

B

Keeping Vertical Siding Vertical

WHEN INSTALLING VERTICAL SIDING, it's a good idea to stop occasionally and make sure the boards are running true and not tilted out of plumb. To do this, measure the distance across the top of the wall to the outside edge of the third cedar board, then take another measurement along the bottom of the wall to the board's edge. The two measurements should be the same. If they're not, make up the difference a little at a time over the next few boards.

For example, if the measurement along the top of the wall is 3⁄8 in. greater than the one at the bottom, leave a 1⁄8-in. gap between each of the next three boards along the bottom of the wall. By making up the discrepancy over a span of three or four boards, the adjustment will be virtually undetectable.

4. Stand the remaining nine pieces of blocking on top of the first girt, making sure you align them with the first row of blocking. Toenail each block to the girt with two 8d nails, then install the remaining girt on top of the second row of blocking, as shown in **B**, nailing it in place with 16d nails. Cut nine more pieces of blocking to 33¾ in. long and nail these final blocks between the second girt and the top wall plate. Refer to the drawing on p. 165 for the location of the rough openings for the two 2-ft. by 4-ft. windows.

Finish the wall

1. Once the wall frame is complete and checked again for square, install two pieces of diagonal metal wind bracing, which will strengthen the wall and help hold it square.

2. Now begin installing the V-jointed, tongue-and-groove cedar siding. Cut the 1×6 siding boards to 89¼ in. (7 in. longer than the wall height). Later, when you stand up the wall, the overhanging siding will cover the 2×6 joist of the floor frame.

Start at the left end of the wall and set the first board flush with the top wall plate and even with the end stud. Make sure the tongue edge faces toward the center of the wall. Fasten the left side of the board (the grooved edge) by nailing through its face with 1½-in. siding nails, driving the nails into the horizontal framing members (i.e., the top and bottom wall plates and girts). Fasten the other side of the board by toenailing at an angle down through the tongue.

3. Set the next board in place alongside the first one, making sure the tongue-and-groove joint fits together tightly. Secure it by nailing through the tongue and into the 2×4 framing.

(You only need to face-nail the first and last boards.) Install the remaining boards in a similar fashion, checking occasionally to make sure the siding isn't tilted out of plumb, as shown in **C**.

Raise the wall

You'll need at least three people to help lift these walls off the floor deck. Be sure to keep children and pets well away from the site during this construction phase.

1. Slide the completed sidewall into position, with its bottom plate about 3 in. from the edge of the floor deck. Now you and your helpers must grasp the wall by the 2×4 top plate and slowly lift up the wall, as shown in **D**. When it's vertical, grab a sledgehammer and move to the outside. Standing at one end of the wall, use the sledgehammer to tap in the very bottom of the wall to ensure it's up against the floor frame.

2. Hold the sledgehammer against the siding while a helper on the inside drives a 3-in. screw down through the bottom plate and into the floor, as shown in **E**.

3. Now move to the center of the wall and repeat the process. Tap in the wall with the sledgehammer so the siding is tight against the floor framing, then hold it in position while it's screwed to the floor. Fasten the remaining end of the wall in a similar manner.

Drive two screws, equally spaced, between each pair of vertical blocking. Angle the screws slightly so they'll go through the wall plate and plywood floor deck and into the 2×6 perimeter joist below. Install a temporary 2×4 diagonal brace on each end of the wall to keep it from falling.

Build the remaining walls

With the first wall in place and temporarily braced with 2×4s to keep it from falling, the next step is to build the frame for the opposite sidewall. It's the same size as the first wall—82¼ in. high by 15 ft. 8 in. long—but it's much easier to build because it doesn't have any windows.

1. Raise the second sidewall just as you did with the first one, secure it to the floor with 3-in. screws, and temporarily brace it in place, as shown in **F**.

2. Now build the rear wall to 9 ft. 5 in. wide. Unlike the sidewalls, this wall has no vertical blocking, but it does have two pieces of diagonal metal wind bracing. It's made up of only six 2×4s: a top plate, a bottom plate, two end studs, and two horizontal girts. Begin by cutting the

2×4 top and bottom wall plates to 9 ft. 5 in. long and the two end studs to 79¼ in. long. Attach the wall plates to the end studs with 16d nails, then check the wall frame for square by comparing the diagonals.

3. Cut two 2×4 girts to 9 ft. 2 in. and set them between the end studs (there's no need to add plywood spacers to the rear or front walls). Nail the first girt 21¼ in. up from the bottom wall plate, driving 16d nails through the end studs and into the girts. Then measure 21¼ in. from the first girt and nail the second girt in place. Use a portable circular saw to let in two pieces of diagonal metal wind bracing.

4. Nail on the siding just as you did for the previous walls. Set it flush with the top plate and allow it to overhang the bottom plate by 7 in.

5. Erect the rear wall, which should fit exactly between the two sidewalls. Screw the bottom plate to the floor deck, as you did with the two sidewalls, then use a 4-ft. level to check both walls in each corner for plumb. Don't worry if the two walls aren't exactly flush; the corner boards will hide any offset. It's more important to make sure the walls are perfectly plumb. Secure each shed corner by screwing through the end stud of the rear wall and into the end stud and blocking of the sidewall, using 3-in. decking screws spaced about 12 in. apart.

6. The front wall is built to the same dimensions as the rear wall: 82¼ in. high by 9 ft. 5 in. wide. Note that this wall has a 58-in.-wide by 72-in.-high rough opening for the door. Cut the top and bottom wall plates to 9 ft. 5 in. long and nail them to the 79¼-in. end studs. Don't

F

worry about the rough opening for now; allow the bottom plate to span the entire length of the wall. It will be cut from the door's threshold after the wall is raised and screwed in place.

7. On each side of the doorway is a 27½-in.-wide section of wall. In each section, install two rows of horizontal girts, spaced 21¼ in. apart, just as you did for the other three walls. When the last wall has been framed and squared, it can be covered with V-jointed cedar siding. However, don't install any siding over the rough doorway opening or over the narrow space directly above the doorway. This 7¼-in.-tall opening is where the transom window will go.

8. After nailing on the siding, lift up the wall between the two sidewalls, tap it into position with the siding held tightly against the floor framing, and screw its bottom plate to the floor deck. Check the corners for plumb, then screw through the end studs of the front wall and into the end studs and blocking of the sidewalls.

Roof Framing

Framing a roof in the traditional manner—with individual rafters, collar ties, and ridge board—is a tedious, time-consuming chore that could easily take all day to complete. However, because you've already assembled the roof trusses, it should take about only 90 minutes to frame the entire roof.

You'll need at least three people for this job: one on the ground and two on top of the sidewalls. The person on the ground is responsible for handing up the trusses and for screwing the trusses to the top wall plate. The two people on top of the sidewalls set the trusses in place and hold them while they're screwed from below.

Start by installing the two gable-end trusses—the ones with the siding attached—to the front and rear walls. Then attach the seven standard trusses to the sidewalls. After the roof is framed, you can nail down the plywood sheathing, then install the shingles.

Set the gable-end trusses

Although standard roof trusses are much lighter and easier to install than gable-end trusses, don't make the mistake of installing them first. If you do, there won't be room to flip the gable-end trusses into place. You'll then have to lift up those heavy trusses from the outside, which is virtually impossible to do without the aid of a small army.

1. Begin by carrying one of the gable-end trusses through the doorway and setting it upside-down on top of the sidewalls, with its peak pointing toward the floor. Slide the truss against the rear wall and check to make sure its exterior surface (the one sided in cedar) is facing in toward the center of the shed.

2. With two helpers standing on ladders situated in opposite corners and close to the rear wall, swing the peak of the truss upward until the helpers on the ladders can reach it. Then stand back while they slowly push the truss to its full, upright position, as shown in **A** on p. 170.

3. With your helpers holding the truss upright, align the 2×4 shoe plate screwed to the bottom of the truss so it's flush with the top wall plate. Secure the truss by driving 3-in. decking screws up through the wall plate and into the shoe; drive one screw every 12 in. Carry in the second gable-end truss and install it over the front wall, using the same procedure.

Install the standard roof trusses

Standard roof trusses are much lighter and easier to install than gable-end trusses are, but you'll still need three people to lift them into position. Set up two ladders inside the shed so two people can work on opposite sidewalls.

1. From outside the shed, hand up the first truss to your helper, who must then pull it onto the wall and slide it across to the helper on the opposite wall. Set the truss directly over the first line of vertical blocking. Working on the left-hand sidewall (the one with the windows), align the very end of the bottom chord on the truss with the outside edge of the wall's top plate.

With each helper standing on the bottom chord to keep the truss from shifting out of position, drive two 3-in. screws up through the top plate and into the bottom of the truss. Because the blocking is directly beneath the truss, you'll have to drive the screws up at a slight angle, placing one screw on each side of the blocking. Don't fasten the other end of the truss just yet.

2. Continue installing trusses in this manner, making sure you place one over every line of vertical blocking, as shown in **B**. Remember: Fasten the trusses to the left-hand sidewall only, aligning the end of each bottom chord with the outer edge of the top plate.

3. After all seven trusses are in position and fastened on one side, you can begin fastening them to the opposite sidewall. This time, start with the truss in the middle of the wall. Using the bottom chord as a guide, push the wall in or out to align the end of the chord with the top wall plate. When the truss and wall are properly aligned, secure the truss with two 3-in. screws driven up through the plate and into the truss. Now attach the remaining six trusses, making sure each bottom chord lines up properly with the top plate.

4. When the trusses are in place, measure the length across the rafter tails from one gable-end truss to the other, and cut two long 2×4s to span the distance. Nail these 2×4s—called subfascia—to the ends of the rafter tails with 3½-in. (16d) nails.

Install the roof sheathing

You'll need nine sheets of ½-in. thick exterior-grade plywood, either BCX or CDX grade, to cover the roof frame. Start with a full 4-ft. by 8-ft. sheet and lay it along the bottom edge of the roof. Align its long, horizontal edge with the 2×4 subfascia nailed to the rafter tails; place its vertical end in the middle of the rafter that forms the gable-end truss.

1. Fasten the plywood along its lower edge with 1½-in. (4d) galvanized nails. Drive one nail through each rafter, making sure the edge of the plywood remains flush with the subfascia, as shown in **C**. If necessary, push the sidewall in or out to align the fascia with the plywood. What you're doing is using the plywood to square up the shed roof. Set another full sheet of plywood end to end with the first one and nail it flush with the subfascia.

2. Use a 4-ft. level to make sure each gable-end truss is plumb. Nail the vertical ends of the plywood to the rafters of each gable-end truss. Check to make sure the rafters are 24 in. o.c., then finish nailing the plywood sheets to the rafters with 1½-in. (4d) galvanized nails spaced about 10 in. apart.

3. You'll need to rip an 8¾-in.-wide strip of plywood to cover the gable overhang at each gable-end truss. Slide the plywood strip into place, making sure it fits snugly between the raised lip of the cedar rake board and the plywood sheet nailed to the rafters, as shown in **D**.

4. Continue to sheathe the rafters on both sides of the roof with plywood. When you get to the top, cut the plywood 1½ in. short of the peak on each side of the roof. That narrow opening will eventually be covered by a ridge vent.

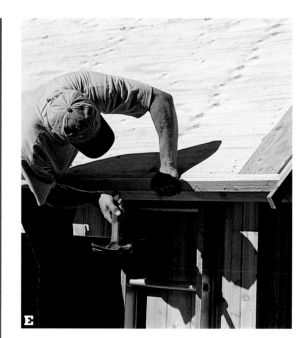

5. Rip a 5-in.-wide fascia board from a cedar 1×6 and nail it to the subfascia with 2½-in.-long ring-shank siding nails. The upper edge of the fascia must be perfectly flush with the top surface of the plywood sheathing.

6. Use a handsaw to trim the rake board on the gable overhang flush with the fascia. Then nail a 1×2 cedar fascia trim to the fascia, as shown in **E**. Again, keep the top edges flush. Cut the overhanging end of the 1×2 rake trim even with the fascia trim.

Roofing

The roof of this Colonial-Style Shed is covered with architectural-style asphalt shingles, also called laminated shingles. Each shingle is 12 in. wide by 36 in. long.

This roof requires three squares (300 sq. ft.) of shingles. For the starter course that covers the ends and edges of the roof, you'll need approxi-

mately 70 linear ft. of standard three-tabs. You'll also need about 18 linear ft. of hip-and-ridge shingles, which are nailed over the ridge vent that runs along the roof peak.

Because each roof plane measures nearly 8 ft. from the drip edge to the peak, you'll have to install three metal roof brackets and a long 2×6 plank to reach the upper portion of the roof.

Install the starter course

Before nailing down the architectural-style shingles, install a starter course of standard three-tab shingles along the sloped ends and lower edges of the roof. This initial course serves two functions: It adds an extra layer of protection to the perimeter of the roof and completely covers the plywood sheathing, so the sheathing won't show through the seams between the architectural-style shingles of the first course.

1. Snap chalklines 11½ in. from the four ends and two edges of the roof. Set the first three-tab shingle on one of the chalklines that runs from the eave to the peak. Make sure the shingle is turned backward, with the tabs facing the center of the roof. That way, there will be solid coverage along the entire perimeter of the roof.

2. Using 1¼-in. roofing nails, begin nailing the starter shingles along the chalkline, as shown in **A**. Secure each shingle with three nails driven through the asphalt sealing strip that runs lengthwise down the center of the shingle. When installed, the 12-in.-wide starter shingles will extend past the roof deck by ½ in., creating a drip-edge overhang. At the peak, trim the shingle flush with the top edge of the plywood; don't cover the ridge-vent space.

3. Repeat this process to install the starter course along the three remaining roof ends. Then nail starter-course shingles along the bottom edge of the roof on both sides of the shed.

Shingle the roof

With the starter course complete, begin installing architectural-style shingles, starting at the eave and working your way up toward the peak. Each shingle must be secured with four 1¼-in. roofing nails.

1. Starting at one end of the roof, begin nailing down the first course of shingles. Set each shingle directly on top of the starter course, making sure it's flush along the drip edge, as shown in **B**. Overlap all vertical seams in the starter course by at least 6 in. Continue laying full-size shingles until you reach the end of the roof, then cut the shingles to length, allowing for the ½-in. overhang at the end.

2. Next, cut 6 in. off a shingle to start the second course. Trimming the first shingle to 30 in. long automatically staggers the seams between the first and second course by 6 in.

Laying Shingles

THERE ARE TWO TECHNIQUES for laying roof shingles. Most pros use the stair-step method, which allows them to install shingles across and up the roof at the same time. The advantage of this method is that you don't have to walk back and forth across the entire roof for each course; instead, you stand in one area. When one section is done, you move down several feet and start the next section.

The second method is more straightforward and, therefore, preferred by most do-it-yourself builders. Called in-line or straight-row shingling, this method consists of completing each course in its entirety from one end of the roof to the other. After each course, you go back to the beginning and start the next one. This method requires a lot more walking back and forth, but fewer mistakes are likely to occur.

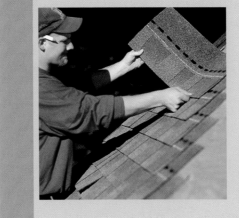

Install the shingles with a 6-in. offset and a 5-in. exposure to the weather. Secure each shingle with four 1¼-in. roofing nails.

Set the first shingle in the second course flush with the end of the roof and align its bottom edge to create a 5-in. exposure to the weather. Continue the course by nailing down full-size shingles.

3. Cut 12 in. off another shingle and start the third course with it. As before, trimming the shingle to 24 in. long will create a 6-in. offset with the seams in the second course. Repeat this pattern as you make your way up the roof. Start the fourth course with an 18-in.-long shingle, the fifth course with a 12-in.-long shingle, and the sixth course with a 6-in.-long shingle. When you reach the seventh course, start with a full-size shingle and begin the pattern all over again.

4. If you maintain a consistent 5-in. exposure to the weather, the last course of shingles will overlap the peak by about 6 in. Use a utility knife to trim the shingles flush with the edge of the plywood roof sheathing, as shown in **C**.

Add the ridge vent

A ridge vent is installed over the narrow space that runs along the roof peak. This simple item is an important part of the shed's ventilation system. Fresh air flows up through the soffit vents—which will be installed under the eaves— and forces air trapped in the shed to exit from the ridge vent.

1. Trim the upper course of shingles flush with the plywood sheathing on both sides of the peak. Next, unroll the 10-in.-wide vent along the center of the ridge, and fasten it to the roof with 2-in.-long roofing nails (or per the manufacturer's directions), as shown in **D**.

2. Drive one nail through the vent and into each rafter. Tap the nails just far enough to hold the vent on the roof; don't compress or crush the fibers. Then, starting at one end and working toward the other, cover the vent with the hip-and-ridge shingles, as shown in **E**. Overlap

Roof Brackets

AFTER YOU'VE SHINGLED ABOUT halfway up the roof, you'll need to stop and install three metal roof brackets so you can safely shingle the upper part of the roof. Place one bracket about 12 in. from each end of the roof and set the third one in the middle. Fasten each bracket with two 16d nails driven through the sheathing and into a rafter. To ensure the brackets are arranged in a straight line, measure down from the top edge of the roof sheathing an equal amount for each bracket. Set the top edge of the brackets on the marks. Set a long, straight 2×8 into the brackets, making sure it fits under the upturned lip at the front of each bracket. Then secure the plank by driving a 1½-in. nail (4d) through the hole in the front of each bracket. Carefully climb onto the plank and continue nailing down shingles, working your way toward the peak.

When it comes time to remove the brackets, start by pulling out the nails that hold the plank in place. Lift the plank out of the brackets and hand it down to a helper on the ground. Next, strike the bottom end of each bracket with a hammer to slide it up the roof about ½ in. That will center each nail head over a large hole in a keyhole slot in the bracket. Pull up to free the brackets.

Gently lift up the shingle and hammer down the nail heads. If you can't lift the shingle high enough to reach the nails, don't force it, or you may crack the shingle. Instead, slide a flat prybar under the shingle and rest it on top of the nail head. Give the bar a couple of good whacks with the hammer to drive the nail home.

After you've shingled halfway up the roof, stop and install three roof brackets. Nail each bracket to a rafter, not just to the plywood.

Lay a sound 2×8 plank across the three roof brackets and continue nailing shingles all the way to the peak.

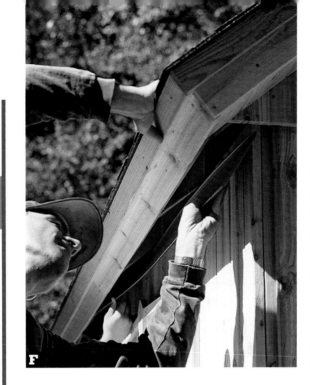

F

the shingles to produce a 5-in. exposure to the weather and fasten each one with two 2-in. roofing nails. Again, be careful not to overdrive the nails and crush the vent.

3. The final step is to install a perforated aluminum soffit vent under the overhanging eave on each side of the shed. The vents are sold in 12-in.-wide sheets that must be cut lengthwise down the middle. Score the sheet a few times with a sharp utility knife, then bend it until it snaps into two 6-in.-wide pieces. Slip the vent under the eave, fitting it between the siding and the fascia board, as shown in **F**. Secure it to each rafter and to the 2×4 subfascia with 1-in. decking screws.

Windows and Exterior Trim

This shed has three windows: two barn sash windows on the left-hand sidewall and a transom window above the pair of swinging doors on the gable end. The transom is a fixed window, meaning it doesn't open. Both 2-ft. by 4-ft. barn sashes tilt in for ventilation.

Barn sash windows are sold at most lumberyards and farm-supply outlets. However, before you can install them, you must trim the exterior window openings with roughsawn cedar and attach an interior window stop. Each sash is held in place by the exterior trim and a 1×2 stop, which is fastened to the sill.

The transom window, which measures 9¾ in. tall by 59 in. long, must be built on site to fit the space above the doors. Fortunately, the window is very easy to make and consists of little more than a wooden frame and two pieces of double-strength glass.

Make the transom window

Cut the parts for the frame of the transom window from leftover 1×6 cedar siding. A tablesaw is the tool of choice for cutting the parts to size, but you can also use a portable circular saw.

1. Start by cutting off the tongue-and-groove edges from two 6-ft. lengths of ¾-in.-thick siding, then rip the boards to the proper width. Note in the drawing on the facing page that the seven vertical mullions and the bottom horizontal rails are 1½ in. wide, but the top rail is 2 in. wide. That's because when the window is installed, the top rail will extend ½ in. behind the siding on the gable-end truss, resulting in a consistent 1½-in.-wide frame. Crosscut the mullions to 9¾ in. long and make the two rails 59 in. long.

2. Next, cut a ³⁄₁₆-in.-deep by ⅞-in.-wide rabbet along the rear edge of the two long rails. These rabbets create a recess for installing the glass panes. You can cut rabbets with a router, a radial-arm saw, or a rabbeting plane, but I prefer to use a tablesaw fitted with a dado blade.

59 in.

1½ in.   1½ in.   1½ in.   1½ in.   1½ in.   8¼ in.   1½ in.   1½ in.

7¾ in.   8¼ in.   8¼ in.   8¼ in.   8¼ in.   7¾ in.

2 in.

Top rail

9¾ in.

1½ in.

Mullion

Bottom rail

7¾-in. × 29¼-in. double-strength glass pane   7¾-in. × 29¼-in. double-strength glass pane

3/16-in. rabbet

Top rail

7/8 in.

9½6-in. × 2-in. rabbet

9¾ in.   7⅞ in.   Double-strength glass

Mullion

7/8 in.

9½6-in. × 1½-in. rabbet

Bottom rail

The narrow transom window installed above the batten doors consists of a wooden frame and two panes of glass. Note that all the frame parts are 1½ in. wide, except the top rail, which is 2 in. wide. Rabbets are milled in the rear of the frame to accept the glass panes.

Also, cut rabbets in both ends of each mullion. These rabbets allow the vertical mullions to fit around the top and bottom rails and lie flat against the glass. However, because the top rail is wider than the bottom rail, the rabbets are different sizes. In the upper end of each mullion, cut a 9/16-in.-deep by 2-in.-wide rabbet. In the lower end, cut each rabbet 9/16 in. deep by 1½ in. wide.

3. Now assemble the frame, using water-resistant carpenter's glue and ¾-in.-long wire nails. Begin by gluing and nailing the two end mullions to the outside surface of the top and bottom rails to form the rectangular frame of the window. Secure both ends of each mullion with two wire nails. Check the frame for square by comparing the diagonals. Then attach the five remaining mullions to the outside surface of the frame, installing the first one 7¾ in. from the end mullion; set the remaining mullions 8¼ in. apart, as shown in the drawing on p. 177.

4. The frame accepts a glass pane that measures 7¾ in. by 58½ in. Although you can order a single piece of glass from a glass shop, I've found out the hard way that it's much easier to transport and install two shorter pieces of glass. Therefore, I suggest getting two pieces of double-strength glass, each measuring 7¾ in. by 29¼ in. To install the glass, place the frame face down on a flat surface and run a thin bead of clear silicone adhesive along the rabbets cut in the rails. Also, run a small bead of silicone on each mullion. Carefully set the glass panes in the frame, then lightly press them into the adhesive. Allow the silicone to cure overnight.

5. To install the transom window, simply raise it above the doorway and slip its top rail behind the siding on the gable-end truss, as shown in **A**. Press the frame's bottom rail against the

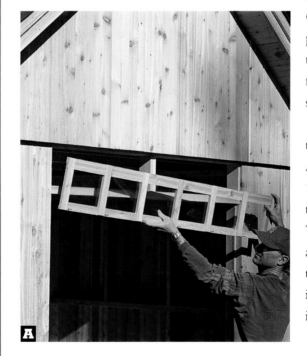

2×4 rough header, then attach the window with ten no. 4, 1½-in. brass screws. Drive five screws, equally spaced, through the bottom rail and five more through the top rail.

Trim the barn-sash windows

The exterior of each 2-ft. by 4-ft. barn-sash window is finished with five pieces of trim, all cut from roughsawn red cedar. The simple pine sashes are available at most lumberyards in various square and rectangular sizes. Installed flat across the bottom of the opening is a windowsill cut from a 5/4 by 6-in. cedar board. On each side of the window is a 3-in.-wide vertical side casing cut from a 1×4. Above the opening is a head casing, which is made from a 1×6 and accented with a couple of 1×2 battens and a decorative keystone. Running horizontally beneath the sill is a 3-in.-wide apron cut from a 1×4.

It's important to note that the head and side casings aren't nailed flush with the edge of the window opening; instead, they extend ⅞ in. into the opening. That creates a lip along the top and sides of the opening to hold the sash in place. On the interior of the rough opening are three pieces of trim. Running lengthwise across the top of the sill is a horizontal window stop cut from a cedar 1×2. It holds the bottom of the sash in place.

Screwed to the wall blocking on each side of the opening is a 12-in.-long vertical 2×2 cleat. A short wooden peg projects from each cleat 1 in. into the opening. When the barn sash is tilted open, it comes to rest against the pegs. The sash itself isn't hinged or screwed in place at all. It simply sits on the windowsill between the window stop and the exterior trim and is held shut with a barrel bolt attached to the interior surface of the sash.

4 in.

3/4 in.

5/4 stock

5 1/2 in.

4 1/8 in.

1 3/8 in.

Keystone

1 1/2 in.

2 in.

12 in.

3/4-in.-dia. hole for wooden peg

Screw and glue the cleat to the wall blocking

Cleat

2 in.

5/4 × 6-in. sill

Bevel the front edge of the sill 10 degrees.

Drip edge (saw kerf)

The exterior of each barn-sash window is trimmed with side casings, a sill, an apron, and a decorative head casing. The exterior trim extends into the opening and creates a lip against which the window closes. A 2×2 cleat and a wooden peg are attached to the inside to hold the sash open.

3/4-in.-dia. wooden peg is set into the cleat; leave 1 in. protruding.

1×2 stop holds sash in place

2×2 cleat

Cedar keystone

Head casing

4 1/8 in.

3 in.

Side casing

23 1/2 in.

48 1/4 in.

46 1/2 in.

Sill

Apron

3 in.

52 1/2 in.

3 in.

3 in.

48 1/8 in.

5 1/2 in.

5/4 × 6-in. windowsill

3 7/8 in.

1 5/8 in.

54 1/8 in.

1. Cut the windowsill to 54⅛ in. long. Saw a 3-in.-wide by 3⅞-in.-deep notch in each end. Set the sill in place on top of the 2×4 rough sill and secure it with six 2½-in.-long siding nails. Next, cut the two vertical side casings to 23½ in. long. Set each one on top of the sill, making sure it overlaps the opening by ⅞ in. Nail the casings to the sidewall.

2. The head casing, similar to the ones found on old barns, comes to a peak in the center, which helps shed water. But rather than just using a single flat board, this casing is dressed up a bit with two 1×2 battens and a center keystone block cut from 5/4 cedar.

Start by cutting a 1×6 to 54⅛ in. long, then rip it to 4⅛ in. wide. To create the pediment shape, measure 3 in. up from each end and make a mark. Draw a line from the top center of the board to the marks on each end. Use a portable circular saw to make angled cuts from the center point to the 3-in. marks. The result is a board that's 3 in. wide at the ends and 4⅛ in. wide in the middle.

3. Make the decorative keystone from a 4-in. by 5½-in. block of 5/4 cedar. Using the drawing on p. 179 as a guide, mark the double-angled shape on the block. The safest way to execute the angled cuts is with a sabersaw or handsaw. The block is a bit too small to cut safely with a power miter saw.

Attach the keystone to the center of the head casing by driving three 1½-in. screws through the back of the casing. Install two 1×2 batten strips on the face of the casing. Cut each one to fit from the keystone to the end of the casing.

4. Attach each batten with three 1½-in. siding nails; be sure to predrill ⅛-in.-dia. pilot holes or

the narrow battens will split. Set the head casing on top of the side casings and secure it with four equally spaced siding nails, as shown in **B**.

5. Make the apron by first cutting a 1×4 to 52½ in. long. Then rip it to 3 in. wide and nail it in place underneath the windowsill, as shown in **C**. Make sure that the ends of the apron are aligned with the outside edge of the side casings.

6. Next, cut a 1×2 to 48⅛ in. long to use as the window stop. To make sure you install the stop in the correct position, use the barn sash as a guide. Set the sash in the opening from the inside and hold it tightly against the lip formed by the exterior trim. Then set the window stop across the sill and press it against the sash. Draw a pencil line along the edge of the stop and onto the sill. Remove the sash, align the stop with the pencil line, and nail it to the sill with four 1½-in. siding nails.

7. The last window-trim pieces to install are the 12-in.-long cleats, which are cut from a 2×2. Bevel-cut the ends at a 45-degree angle, then measure 2 in. down from one end and bore a ¾-in.-dia. by 1-in.-deep hole in the side of each cleat. Cut off two 2-in.-long pegs from a length of ¾-in.-dia. hardwood dowel rod. Glue the pegs in the holes.

Hold each cleat against the vertical 2×4 blocking on each side of the window opening. The pegs must face the opening and the bottom ends of the cleats must be flush with the top of

the 2×4 rough sill. Secure each cleat with two 3-in. screws, as shown in **D**.

8. Next, mount a 2-in. sliding barrel bolt on the top center of the sash. Slip the sash into place from the inside, as shown in **E**. Finally, bore a 5/16-in.-dia. hole for the barrel bolt in the 2×4 rough header above the sash.

Install the corner boards

The cedar corner boards that cover the four corners of the shed are the last pieces of exterior trim to be installed. You can cut and install these eight boards one at a time, but it's much easier to preassemble a corner cap for each shed corner by first nailing together two boards.

1. Each 90-in.-long corner cap is made from a roughsawn cedar 1×4 and 1×6. Rip the 1×6 to 4¼ in. wide, then nail it to the narrower 1×4 with 2½-in.-long siding nails spaced 12 in. apart. This creates an L-shaped corner cap measuring 4¼ in. square.

Building Batten Doors

THE TRADITIONAL BATTEN DOORS on this shed are made from the same V-jointed, tongue-and-groove cedar boards used for the siding. The boards in each door are held together by three horizontal battens and two diagonal battens, which are screwed across the back. The 4¼-in.-wide battens are also cut from the V-jointed cedar boards. This particular style is called a Z-batten door because the diagonal battens form a Z-shaped pattern. Each door measures 29¼ in. wide by 74 in. high. Here are the basic steps for building each one:

1. Cut six tongue-and-groove 1×6s to 74 in. long. Use a tablesaw to rip the tongue off one board and the groove off another.

2. Lay the boards on a workbench with their roughsawn surfaces facing down. Align the ends flush, then use pipe clamps to draw the boards together into a single panel, as shown in **A**. Make sure the first and last boards are square at the edges.

3. Cut three horizontal rails from a 4¼-in.-wide cedar board; make each one 27⅞ in. long. Apply construction adhesive to the backs of the rails, as shown in **B**, then attach them with 1¼-in. decking screws, as shown in **C**.

4. Cut a Z-batten to fit diagonally from the bottom rail to the center rail. Run a bead of adhesive across the back and set it in place. Cut a second Z-batten to fit between the center rail and the upper rail. Run this one in the opposite diagonal direction of the lower Z-batten, as shown in **D**.

5. Screw a 2-in.-wide astragal strip to the latch side of the left-hand door, as shown in **E**. This overhangs the door edge by 1 in. to form a support strip for the right-hand door to close against.

6. Hang each door with a pair of 12-in. strap hinges. Bolt the strap portion to the door with carriage bolts, then set the door in the opening. Use shims to create the proper clearance at the top and bottom of the door.

7. Slip a cast-iron hinge cup onto the top and bottom of the hinge pin protruding from each strap, as shown in **F**. Fasten the cups to the shed wall with the screws provided.

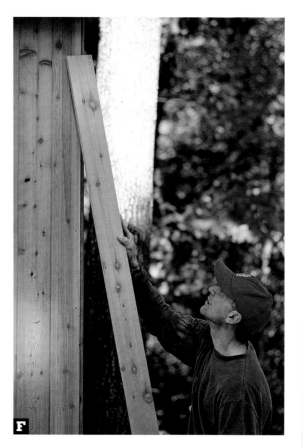

2. Slip the corner cap into place, making sure it fits tightly against the soffit vent on the sidewalls and under the siding on the gable-end truss on the front and rear walls, as shown in **F** on p. 183. The bottom end of the corner boards should be flush with the siding. Secure the corner cap to the shed with 2½-in.-long siding nails spaced 12 in. apart.

PRO TIP

Make sure the lumber you use to build the batten doors is thoroughly dry. Otherwise, the doors will shrink when the lumber dries.

WHAT CAN GO WRONG

Over time, shed doors tend to sag and rub together, making it difficult to close them. This is often caused when hinge screws become loose or strip out their holes. To remedy this problem, start by removing the hinge-cup screws and shimming the door back into its original position. Reattach the hinge cups with longer screws. Drive the screws into solid wall framing, not just the siding.

DESIGN
OPTIONS
COLONIAL SHED VARIATIONS

▲ There's no shortage of storage in this spacious 12-ft. by 20-ft. Colonial-style barn. It features vertical-board siding, sliding door, cupola, and cobblestone ramp.

▶ This 12-ft. by 16-ft. blue-stained garden shed has two doublewide doors, which provide easy access to the interior. It also has three tilt-in barn sash windows, a transom window, and a gable roof covered with architectural-style shingles.

This charming backyard building features cedar clapboards on the sidewall, board-and-batten siding on the gable end wall, and asphalt shingles on the roof. It measures 12 ft. by 14 ft., which is an ideal size for use as a storage building, garden shed, or kid's playhouse.

The rustic, natural beauty of red cedar is on display in this attractive 10-ft. by 14-ft. workshop. Cedar shingles cover the sidewalls; the roof is made of cedar shakes. The gable wall is sided with cedar board-and-batten siding.

Built as a tractor barn, this 12-ft. by 22-ft. outbuilding features vertical-board siding, standing-seam metal roof, and doublewide doors salvaged from an old dairy barn. The upper window on the gable end wall provides access to an overhead storage loft.

GAMBREL STORAGE BARN

Traditional barn architecture and modern building methods come together in this spacious 12-ft. by 20-ft. storage barn, which features a gambrel-style roof, grooved plywood siding, aluminum windows, and a pair of sliding doors. At 240 sq. ft., the building is supported by a pole-barn foundation consisting of six 4×4 treated posts set on concrete footings. Note that the entire structure is suspended from the posts; there's no wooden floor.

The ground inside the barn is covered with processed stone, creating a floor that's just about even with the ground outside. That makes it easy to store lawn tractors, farm machinery, and trailers. Another benefit is that the gambrel roof creates a spacious storage loft above the ceiling joists. Need one more reason to build this barn? Consider this: It'll make the neighbors really jealous. (To order plans for the Gambrel Storage Barn, see Resources on p. 214.) ▶ ▶ ▶

Site Prep and Footings

The local building code required a permanent, frost-proof foundation for this storage barn because it's more than 200 sq. ft. Codes differ from region to region, but chances are that you, too, will need to build a frost-proof foundation. If you're planning to use the barn as a woodshop or carriage house, consider pouring a monolithic concrete-slab foundation. For general storage, the pole-barn foundation used on this barn works fine—and it's a whole lot easier to build.

Remove the sod

Before you can start laying out the foundation, you must remove the sod from the building site. That not only creates a more stable surface for the processed-stone floor, but it also gives you the opportunity to rake the ground flat and somewhat level. Note that it's necessary to remove the sod from an area slightly larger than the footprint of the building; in this case, you must clear a 14-ft. by 22-ft. area. That will create a well-draining, grass-free buffer zone around the perimeter of the barn.

There are two basic ways to remove sod: by hand with a flat-blade shovel or with a gas-powered sod cutter. The choice is easy, considering that you must clear an area that's more than 300 sq. ft.: Go out and rent a sod cutter.

1. Mark out the 14-ft. by 22-ft. area using spray paint or white flour sprinkled from a can.

2. Run the sod cutter back and forth from one end of the building site to the other. Have a helper roll up and remove the sod strips, as shown in .

Set up the batter boards

Once the sod is removed, the next step is to determine exactly where to dig the six footing holes. One of the easiest and most accurate methods for laying out footing holes is to use strings and batter boards. These pairs of vertical stakes are placed a few feet beyond each corner of the foundation, then strings are stretched tightly between them to represent the rectangular outline of the foundation. Key dimensions are then taken off the strings with a tape measure and plumb bob.

1. Using the drawing on p. 191 as a guide, install the first batter board near one corner of the foundation. Drive in the two vertical stakes, spacing them about 3 ft. apart, then screw a crosspiece to the stakes, making sure it's level, as shown in **B** on p. 190.

Tack a small nail in the top edge of the crosspiece centered between the stakes; leave about 1 in. of the nail protruding. Install the opposite batter board in a similar manner, but don't attach the crosspiece just yet. Tie a string to the nail on the first batter board and pull it across to the second one.

Architectural-style roof shingles

Plywood roof sheathing

Ridge vent

2×4 roof rafter

Plywood gusset

Fascia trim

Fascia board

Rake trim

Rake board

Soffit vent

Cedar corner board

Grooved plywood siding

Aluminum sliding window

Faux board-and-batten doors

2×10 skirt board

2×4 wall stud

2×6 bottom chord

6×6 grade beam

Pole-barn foundation

Use nylon mason's line,
not cotton string, to lay
out the batter boards.
The nylon line is very
strong, stretches taut,
and doesn't twist and
tangle when rewound.

**When measuring long
distances,** have a helper
support the middle of the
tape measure to prevent
it from sagging.

**To easily locate
plumb-bob marks** on
the ground, push a nail
through a slip of brightly
colored paper and press
it into the ground on the
center mark.

2. Hang a line level on the string and raise or
lower the crosspiece of the second batter board
until the string is level, as shown in **C**. Screw
the crosspiece to the stakes and drive a nail in
the top edge. Pull the string taut and tie it off.
Repeat this procedure to install the remaining
three pairs of batter boards.

3. After stretching all four strings taut between
the batter boards, check them for square by
holding up a framing square where the strings

intersect at 90-degree angles. The corner is
square when the strings align with the edges of
the framing square. Now, check for square once
again, but this time measure the diagonals of
the rectangle formed by the four strings (you'll
need a helper for this). When the two measurements are equal, the layout strings are square.

4. Note from the drawing on the facing page
that the 144-in.-wide by 236-in.-long rectangle
formed by the strings represents the outer edge
of the 4×4 posts. To find the center of the footing holes (and the posts), measure in from the
string 1¾ in., which is half the width of the
3½-in.-sq. posts. Hold the end of the tape measure against the string, making sure you don't
deflect the string.

5. Line up the plumb-bob line with the
1¾-in. graduation on the tape measure and
drop the bob to mark the ground, as shown
in **D**. The mark represents the center of the
12-in.-dia. footing hole and 4×4 post. Use this
same technique to mark the locations of the
remaining five holes.

Dig and pour
the concrete footings

Each of the six footing holes must be lined with a 12-in.-dia. fiber-form tube. The tubes come in 12-ft.-long sections and are sold at most lumberyards and home centers.

1. Use a handsaw or a sabersaw to cut the tube into pieces 6 in. longer than the depth of the hole.

Stand a fiber-form tube over one of the footing-hole center marks made by the plumb bob. Peer down inside the tube and center it directly over the mark, then score the ground around the perimeter of the tube with a shovel or mark it with spray paint.

2. Dig the hole to the desired depth, using both a shovel and a post-hole digger. Lower the fiber-form tube into the hole and backfill around it, as shown in **E**.

FOUNDATION PLAN

19 ft. 8 in.

4×4 post

Batter board

1¾ in.

12-in.-dia. footing hole

12 ft. 0 in.

90-degree angle

90-degree angle

Strings pulled taut between batter boards

Batter boards and strings are used to locate the six footing holes. The strings represent the outside surface of each 4×4 post. Therefore, you must measure in from the strings 1¾ in. (half the thickness of the 4×4s) to find the center of the 12-in.-dia. footing holes.

ACCORDING TO CODE

When building the foundation, be sure to dig the footing holes down to the frost line, but don't pour the concrete until after the building inspector has had a chance to inspect the holes. That's the building department's only assurance that the holes are dug to the proper diameter and depth.

PROTIP

If you can't dig the footing holes with a shovel or post-hole digger in ground that's too hard or rocky, an excavator can bore the holes with a power auger.

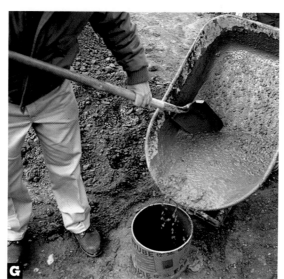

Be careful not to hit the protruding tube with the lip of the wheelbarrow or shovel, or you may knock it out of position. Pour the rest of the concrete footings, using one bag of concrete per hole. Allow the concrete to cure for at least 24 hours before setting the posts.

Wall Framing

The wall system of this pole barn consists of four 4×4 corner posts, two 4×4 intermediate posts, four sidewall sections, and two gable-end walls. The six posts are set on top of the concrete footings. The wall frames are assembled from 2×4s and plywood siding, then set between the posts. Two sidewall sections are needed to form each side of the 20-ft.-long barn. Also, there are a total of five window openings: one in each sidewall section and another in the rear gable-end wall.

Install the posts

Each of the six posts is made from a 12-ft.-long pressure-treated 4×4. Nailed to two sides of each post are shorter 2×4 pressure-treated cleats. Later, the walls will be set between the posts and on top of the cleats.

3. As you're backfilling, have a helper hold a 2-ft. level against the inside of the tube to make sure it remains plumb, as shown in **F**. Dig the remaining five footing holes and install the fiber-form tubes.

4. Next, dump a bag of ready-mix concrete into a wheelbarrow, add water according to the directions, and blend it smooth with a hoe. Pour the entire contents of the wheelbarrow into one of the tubes to form a footing at the bottom of the hole, as shown in **G**.

1. Start by making the two intermediate posts, which will be installed in the middle of the sidewalls. Cut 2×4 cleats to extend from the bottom end of the posts to about 12 in. above the finished grade, which will be the ground level at the doorway when the barn is completed.

2. Using 3½-in. (16d) galvanized nails, attach the 2×4 cleats to opposite sides of each intermediate post, as shown in **A**, but don't nail within 12 in. of the top of the cleat (that part will be trimmed off later). Set the intermediate posts in the footing holes in the middle of the sidewalls.

Set the corner posts

1. Next, nail two 2×4 cleats to each corner post, but this time fasten them to adjacent sides, not opposite sides, of the post. This is necessary because the walls meet the corner posts at a 90-degree angle. Set the corner posts in the footing holes, again making sure the cleats extend from the bottom end of the posts to about 12 in. above the finished grade. Temporarily screw a long 2×4 across the three posts along each side of the foundation to

hold them in place, but don't backfill around the posts just yet.

2. Find the post situated at the highest corner on the site and make a mark on the cleat 9 in. above the finished grade. Using that mark as a reference point, stretch a mason's string to the post at the opposite corner. Use a line level to level the string, then mark the cleats at each post where the string crosses the cleats. Repeat this procedure to mark the cleats on the remaining posts. Lift the posts out of the holes and use a circular saw to trim the cleats to the lines.

3. You'll also need to add one additional cleat to the inside surfaces of each of the two corner posts on the gable-end wall that will be framed for the sliding doors. These two extra cleats support the 6×6 grade beam installed across the threshold of the doorway. Measure 8½ in. down from the top of the first cleat nailed to the inside of each corner post and make a mark. Measure from the bottom of the post to the mark and cut a 2×4 cleat to fit. Nail this second cleat to the first cleat with 3½-in. (16d) galvanized nails. Then set all the posts back in their respective holes and temporarily screw a long 2×4 across them to hold them in place.

4. Cut a pressure-treated 2×10 to 11 ft. 5 in., which is the exact width of the rear wall, and nail it to the two corner posts at the rear of the foundation, as shown in **B** on p. 194. Hold the top edge of the 2×10 skirtboard 1½ in. above the 2×4 cleats nailed to the post, making sure it is flush with the bottom wall plate.

5. At the front of the foundation, where the sliding door will eventually be hung, install

TRADE SECRET

The tops of all the 2×4 cleats must be at the same height: 9 in. above the finished grade. The easiest way to do this is to cut the cleats about 12 in. too long, then attach them to the posts without nails in the top 12 in. or so. After the posts are in place, stretch a string across them and use a line level to level the string. Mark the cleats where indicated. Lift out the posts and cut the cleats at the marks.

a pressure-treated 6×6 grade beam to serve as a threshold between the two corner posts, as shown in **C**. Lay the timber on a bed of compacted processed stone. Make sure each end of the beam rests on the 2×4 cleats, then fasten the 6×6 beam to the posts with 3-in. decking screws.

Frame the walls

The barn's walls are framed with 2×4s and sheathed with grooved plywood siding. This

is done while the walls lie flat on the ground; then they're lifted up and set between the posts.

1. Begin by building the four sections that make up the two long sidewalls. For each section, cut a top and a bottom wall plate to 9 ft. 4¾ in. long. Next, cut six 2×4 studs to 74½ in. long.

2. Use 3-in. decking screws to fasten the wall plates to the two end studs to create the rectangular wall frame. Check the frame for square by measuring the diagonals. Then install the remaining four studs and frame out the rough opening for the 2-ft. by 3-ft. aluminum window, as shown in **D**, following the drawing on the facing page.

When building the walls, lay out the studs for each one by starting from the same end. That way, the studs in one wall will align perfectly with the studs in the opposite wall. This is important because the studs will later be used to position the roof trusses.

22¼ in.   24 in.   24 in.   24 in.   18½ in.

11 in.

4½ in.

26¼-in. × 36¼-in.
rough opening

This end
faces the
corner post.

26¼ in.

74½ in.

40¾ in.

This end
faces the
adjacent
wall frame.

9 ft. 4¾ in.

Each long side of the barn is made from two 9-ft.-4¾-in.-long wall frames,
which are built entirely out of 2×4s. Note that each frame has a rough opening
for a 2-ft. by 3-ft. window.

3. Now prepare to cover the wall frame with
⅝-in.-thick grooved plywood siding. First,
crosscut a plywood sheet to 82 in. long, which
is 4½ in. longer than the wall height. Later,
when the wall is set between the posts, the
overhanging plywood will partially conceal
the 2×10 skirtboard nailed across the inside
of the posts. Apply a continuous bead of construction adhesive to the 2×4 wall frame. Lay
the plywood on the frame, making sure its
top end is flush with the top wall plate and
its vertical edge overlaps the end stud by 3½ in.
That extra bit of siding will cover the 4×4
corner post.

4. Nail the plywood to the wall frame with
1½-in. ring-shank siding nails spaced 10 in.
apart, as shown in **E** on p. 196. Finish installing
the siding, making sure you cut it flush with the
edges of the rough window opening.

The last piece of siding extends past the end
stud, too, but by only 1¾ in., which is half the
width of the intermediate post. Now build three
more identical sidewall sections. Make each one
77½ in. tall by 117¾ in. long, remembering to
frame in the window in each section.

PRO TIP

If you rub wax or bar soap onto screw threads, they'll spin into the wood with much less strain on you and the drill's motor.

5. Next, build the two gable-end walls, making each one 77½ in. tall by 137 in. long. Note that the rear wall has a 25¼-in. by 36¼-in. rough window opening and the front wall has a 66-in. by 73-in. rough door opening. For each wall, cut a top and a bottom wall plate to 11 ft. 5 in. long. Then cut six 2×4 studs to 74½ in. long. Fasten the wall plates to the studs with 3-in. decking screws. Space the studs 24 in. o.c. Cover the two gable-end walls with plywood siding. Make sure the plywood extends past both ends of each gable-end wall by 3½ in. to effectively cover the 4×4 corner posts. You're now ready to install the walls, starting at the rear of the barn.

Raise the walls

With the help of a friend, carry the rear wall to the back of the foundation. Slide it between the two rear posts, making sure its bottom wall plate is on top of the 2×4 cleats nailed to the posts.

1. Raise the wall to a full upright position, as shown in **F**; if the fit is too tight, push out on the posts.

2. As your helper holds the wall in place, move to the inside and fasten it to the posts with 3-in. decking screws spaced about 12 in. apart, as shown in **G**.

3. Next, install the first of the four sidewalls. These sections are much lighter and easier to handle than the rear wall. Slide the first one between the rear corner post and the intermediate post, as shown in **H**.

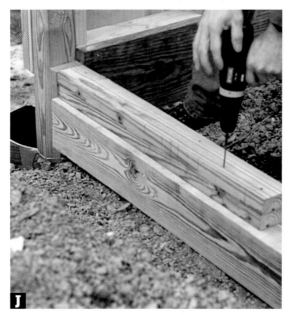

Again, make sure its bottom plate sits directly on top of the 2×4 cleats. As your helper pushes the wall tightly against the posts from the outside, secure the wall by screwing through the end studs, as shown in **I**. The wall is in the proper position when the end studs completely cover the sides of the post.

4. Install the second sidewall next to the one you just screwed in place. Check to make sure the plywood seam between the two wall sections fits together tightly at the intermediate post. Don't worry if the opposite end of the wall doesn't fully cover the corner post; that joint will be concealed later by corner boards. It's more important to create a tight seam at the intermediate post. After screwing the second sidewall in place, move to the opposite side of the foundation and install the two remaining sidewall sections.

Erect the front wall

The front wall is more than 11 ft. long, but it's very easy to lift into place because of the large doorway opening. However, before you can install it, you must attach double 2×4 sill plates to

the 6×6 grade beam on each side of the doorway opening. These boards are required because the beam is 3 in. lower than the cleats that support the sidewalls. The sill plates will raise the front wall even with the sidewalls.

1. Cut four 34-in.-long sill plates from pressure-treated 2×4s. Stack two sill plates on top of the 6×6; make sure they're butted tightly against the corner post and centered on the 6×6. Fasten each plate with 3-in. decking screws, as shown in **J** on p. 197 Fasten the other double sill plate to the opposite end of the grade beam.

2. Set the front wall between the corner posts and tilt it up into place, as shown in **K**.

3. Screw through the end studs and into the posts, just as you did for the other walls, but also screw down through the wall's bottom plate and into the double sill plate, as shown in **L**. Move to the outside of the barn and fasten the plywood siding to each of the six posts with 1½-in. ring-shank siding nails spaced 10 in. to 12 in. apart.

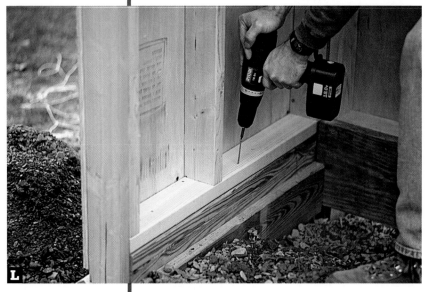

Complete the wall framing

Finish up the wall framing by nailing a 2×10 skirtboard to the inside surface of the two sidewalls.

1. Hold the 2×10 even with the top of the wall's bottom plate and drive 3½-in. (16d) galvanized nails into the edge of the bottom plate and into the 4×4 posts, as shown in **M**. The skirtboards add structural integrity to the walls and create a perimeter frame, which holds the processed stone in place.

2. The next step is to encase the six 4×4 posts in concrete to permanently hold them in position. Don't forget that at this stage the posts are just sitting on top of concrete footings; they're not secured within the fiber-form tubes. However, before you can encase them, you must make sure the walls are straight; otherwise, you'll never get the roof trusses to fit.

To straighten the walls, start by temporarily screwing a short 2×4 block to each end of the

4. After you've confirmed that all the walls are straight, mix up a bag of concrete in a wheelbarrow. Use a sharp utility knife to cut the fiber-form tube flush with the ground, then shovel one-half of the concrete into the tube, as shown in , making sure you distribute it evenly on all sides of the post. Repeat the process for the other five posts, using half a bag of concrete per hole. Allow the concrete to cure overnight, then fill each hole to the very top with processed stone or gravel.

Roof Framing

It's never easy to frame a roof as big as this one, but using prebuilt trusses makes the job much simpler. Once you build the two gable-end trusses, it should take about two hours or so to frame the roof and another couple of hours to nail down the plywood roof sheathing.

You'll need at least four people to frame this roof: two on the ground handing up the trusses and two on top of the sidewalls setting the trusses in place. The ground workers also screw the trusses to the top wall plates from below. Note that it's important to place one truss directly over each wall stud. That way, the wall frames, posts, and footings will properly support the weight of the roof.

Roof-truss assembly

A typical gable roof truss has only three major parts: two angled rafters and a horizontal bottom chord. Each of the nine standard trusses in this gambrel barn is framed with seven pieces: four rafters, two vertical braces, and one bottom chord. The top two rafters meet at the peak and form a relatively shallow 6-in-12 slope. The roofline then breaks down sharply where the second pair of rafters forms a very steep 24-in-12 slope.

sidewall near the bottom of the wall. Stretch a very taut string across the blocks and tie it off.

3. Using a third 2×4 block as a measuring gauge, slip the block behind the string at several places along the wall, as shown in N. If there's not enough space for the block to slide behind the string, move the wall in toward the center of the barn. If the space between the string and the siding is greater than the thickness of the 2×4 block, push the wall out.

2×4 rafter, 60 in. long

27-degree plumb cut

27-degree cut

54-degree cut

½-in. plywood knee gusset (both sides)

½-in. plywood side gusset (both sides)

½-in. plywood ridge gusset (both sides)

2×4 rafter, 68 in. long

2×4 brace, 50¾ in. long

Plywood stop blocks

½-in. plywood simple gusset (both sides)

2×6 bottom chord, 12 ft. long

Gambrel roof trusses are slightly more difficult to build than standard gable trusses. Each one has four 2×4 rafters, two 2×4 vertical braces, and one 2×6 bottom chord. One truss is assembled on the ground and used as a template for the remaining trusses. A series of plywood stop blocks are screwed to the template truss and used to hold the parts in proper alignment. Plywood gusset plates are glued and nailed across the joints to hold each truss together.

The rafters and braces are cut from 2×4s, but for added strength the bottom chord is a 2×6. The boards that make up the trusses are held together with plywood gusset plates glued and nailed to each side of the truss.

1. For each truss, cut two 2×4 top rafters to 60 in. long. Trim one end of each rafter at a 27-degree angle and cut the other end square. Then cut two 2×4 lower rafters to 68 in.; miter-cut one end at a 54-degree angle. Next, cut the two vertical 2×4 braces to 50¾ in. Again, trim one end at a 27-degree

angle and cut the other end square. Finally, cut the 2×6 bottom chord to 12 ft. long, mitering both ends at a 27-degree angle.

2. For each truss, you must also cut 14 gusset plates from ½-in. ACX plywood. Cut the two triangular ridge gussets to 9½ in. high by 24 in. wide, and cut eight side gussets to 13 in. by 18 in. In addition, you'll need four 3½-in.-wide by 12-in.-long simple gussets to secure the vertical braces to the bottom chord. Next, cut twenty-five 3-in. by 6-in. stop blocks from ½-in. plywood.

5. Lay the parts for the next truss directly on top of the template truss, as shown in **B**. The stop blocks will hold each piece in place, but check to make sure the joints fit together tightly.

6. Glue and nail on the gusset plates with construction adhesive and 1-in. roofing nails, as shown in **C**. Flip over the truss and attach gussets to the other side. After assembling the last truss, add gussets to the template truss and unscrew the stop blocks.

Build the gable-end trusses

In addition to the nine standard trusses, you'll need to build two gable-end trusses.

1. Start by assembling two trusses to the same dimensions as the standard trusses. Next, add five more vertical 2×4 braces, spacing them equally between the two original braces. These extra pieces provide solid support for the grooved plywood siding. Also, screw a 10-ft.-long 2×4 shoe plate to the bottom chord. When the truss is installed, the shoe plate will rest on top of the gable-end wall.

3. Assemble the parts to make one truss, but attach gusset plates to only one side, using construction adhesive and 1-in. roofing nails. Use 10 nails per gusset.

4. Temporarily fasten the stop blocks to the edges of the rafters, braces, and bottom chord with 1¼-in. screws, as shown in **A**. Refer to the drawing on the facing page for the exact placement of the stop blocks. You can now use this truss as a template for assembling the others.

Assembling the trusses on the ground significantly reduces the amount of time it takes to frame the roof. However, you have to make sure you build all the trusses to exactly the same size. If one truss is even slightly taller than the others, it'll affect the installation of many other components, including the roof sheathing and the subfascia.

Glue and nail plywood gusset plates to only one side of the truss, then flip over the truss and sheathe it with plywood siding. Make sure each seam between the plywood pieces fall on the center of a 2×4 brace.

2. Secure the plywood with construction adhesive and 1½-in. ring-shank siding nails, as shown in **D**.

3. Run the siding long, then use a circular saw to trim the plywood flush with the rafters, as shown in **E**.

4. Next, make the rake boards by ripping roughsawn cedar 1×6s to 5 in. wide. Miter the rakes to match the roof slope and cut them to extend about 6 in. past the rafter tail. That extra bit of rake will be trimmed off after the fascia is installed. Position them to project ½ in. above the top edge of the rafters; that raised lip will hide the edge of the plywood roof sheathing.

5. Secure the rake boards to the truss with 1½-in. siding nails, as shown in **F**. Finally, nail 1×2 cedar boards—called rake trim—to the rake boards.

6. Carry one of the gable-end trusses inside the barn and set it upside-down on top of the sidewalls. Make sure its plywood surface faces toward the center of the barn. Grab the peak of the truss and rotate it upward. Have two helpers on ladders slowly push the truss to a full, upright position, as shown in **G**.

7. Align the 2×4 shoe plate on the truss with the top wall plate. Secure the truss by driving 3-in. decking screws down through the shoe

SAFETY FIRST

When framing a roof, it's a good idea to nail a temporary 2×4 brace across the trusses to hold them upright and prevent them from accidentally falling over. Set the brace horizontally across the trusses, positioning it near the upper end of the lower rafters. Nail the end of the brace to the gable-end truss. As you install each truss, secure it by nailing through the brace and into the rafter.

PRO TIP

Glue an empty tuna can to the top of your ladder. Then place a magnet in the can and use it to hold nails and screws

TRADE SECRET

When it comes time to install the roof trusses, employ a tag-team approach: Have one person stay up on the wall and set a truss directly over a wall stud. Then, as you drive two screws from below, have him or her stand on top of the bottom chord to prevent the truss from sliding out of position. Note that it's necessary to drive the screws at a slight angle in order to miss the stud.

and into the top wall plate; drive one screw every 12 in., as shown in **H**. Install the second gable-end truss in a similar manner.

Install the roof trusses

To install the nine standard roof trusses, you must lift each one onto the walls from the outside. Fortunately, the trusses aren't all that heavy, but at 7 ft. tall by 12 ft. long, they're a little unwieldy. Enlist the help of at least four people: Place two workers inside on ladders and two outside to handle the trusses.

1. Lift up the first truss, set it on top of the sidewall, and slide it across to the helper at the opposite wall, as shown in **I**.

2. Set the truss directly over the first wall stud. Line up the very end of the bottom chord with the outside edge of the wall's top plate. Drive two 3-in. screws up through the top plate and into the bottom of the truss, as shown in **J** on p. 204. Don't fasten the other end of the truss just yet. Install the remaining trusses, making sure you place one over each stud.

Mark the 2×4 with the location of the rafters, which are spaced 24 in. o.c. Hold the 2×4 subfascia against the ends of the rafter tails, align each tail with the layout mark, and nail the subfascia to the ends of the tails with 3½-in. (16d) nails, as shown in **K**.

Sheathe the roof

You'll need 15 sheets of ½-in.-thick exterior-grade plywood (BCX or CDX grade) to cover this gambrel roof. Sheathe the upper roof slope first, then cover the lower slope.

1. Start with a full 4-ft. by 8-ft. sheet, laying it on top of the rafters and aligning its lower edge flush with the 2×4 subfascia. Butt the end of the sheet tightly against the rake board, then fasten the plywood to the rafters and subfascia with 1½-in. (4d) nails spaced 10 in. apart.

Install another full sheet of plywood, followed by a half sheet. At the peak, rip the plywood to fit, leaving it 1½ in. short of the peak on each side of the roof; that narrow space will be covered later by the ridge vent.

2. When the upper roof slope is covered, start sheathing the lower slope, beginning with a full 4-ft. by 8-ft. sheet of plywood. Slide the sheet over the rafters and fit it tightly underneath the overhanging rafter tails of the upper roof slope, as shown in **L**. Continue covering the upper portion of the lower roof slope with plywood, but don't sheathe the lower section just yet. You must install the soffit vent first.

3. Nail a 2×4 subfascia to the ends of the rafter tails, then slip a 6-in.-wide section of perforated aluminum soffit vent underneath the roof overhang and press it tightly against the underside of the rafter tails.

3. Now begin fastening the trusses to the opposite sidewall, starting with the truss in the middle. Using the bottom chord as a guide, push the wall in or out to align the end of the chord with the top wall plate, then fasten the truss with two 3-in. screws driven from below. With the middle truss holding the wall straight, attach the remaining trusses.

4. Next, cut a long 2×4 subfascia to span the distance across the tails of the upper rafters.

TRADE SECRET

There's a detail on this gambrel roof that you seldom see on outbuildings. It has a fascia board installed along the edge of the upper roof slope. Ordinarily, the roof shingles are simply bent to cover the joint between the upper and the lower slopes. That not only looks bad but also creates a fault line along the joint, where the shingles will be more likely to crack and wear out.

about 12 in. apart. When the vent is installed, you can finish nailing down the plywood roof sheathing,

Add the fascia

1. Cut 5-in.-wide fascia boards from rough-sawn cedar 1×6s and nail them to the subfascia with 2½-in.-long siding nails, as shown in **N**. Make sure the upper edge of the fascia is even with the plywood sheathing.

Note that you'll need to cut two fascia boards to span the distance. Join the two boards together with a simple butt joint. Drill ⅛-in.-dia. pilot holes to prevent the nails from splitting the ends of the boards.

2. Use a handsaw to trim the rake board on the gable-end truss flush with the fascia, as shown in **O** on p. 206.

3. Nail a 1×2 cedar fascia trim to the fascia. Again, keep the top edges flush, as shown in **P** on p. 206. Then trim the overhanging end of the 1×2 rake trim even with the fascia trim.

4. Use a length of aluminum drip-edge flashing to hold the bottom edge of the vent to the subfascia. Place the L-shaped flashing over the vent and fasten it to the subfascia with 1-in. roofing nails, as shown in **M**. The reason it's necessary to use flashing is that there's not enough room below the rafter tails to nail or screw the vent in place.

5. Fasten the top edge of the soffit vent to the plywood siding with 1-in. roofing nails spaced

PRO**TIP**

The best time to begin shingling the roof is in the early morning or late afternoon, when the weather is a little cooler.

PRO**TIP**

Instead of using a utility knife to cut shingles, rent a manual roof-shingle cutter. It operates like a giant paper cutter and slices shingles quickly and safely.

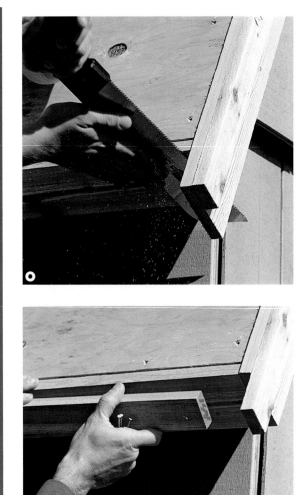

Roofing

You'll need four and half squares (450 sq. ft.)—about 18 bundles—of architectural-style asphalt shingles to cover the roof of this Gambrel Storage Barn. You'll also need 130 linear ft. of three-tab shingles (four bundles) for the starter course and 20 linear ft. of hip-and-ridge shingles (one bundle) to cover the ridge vent.

Install the starter course

The starter course consists of three-tab shingles nailed along the ends and edges of the roof. This course protects the perimeter of the roof and ensures that the plywood sheathing doesn't show between the architectural shingles. To completely cover the plywood, you must install the starter-course shingles upside-down, with the tabs facing toward the center of the roof.

1. Start by snapping a chalkline 11½ in. from the bottom edge and each end of the roof. Because the shingles are 12 in. wide, this will produce a ½-in. drip-edge overhang. Set a three-tab shingle on one of the vertical chalklines that runs up to the peak, fastening the shingle with three 1¼-in. roofing nails driven through the asphalt sealing strip. Install another shingle vertically on the opposite end of the roof.

2. Install starter-course shingles along the bottom edge of the roof, setting them on the chalkline and securing each with three nails, as shown in **A**. When you reach the end of the roof, cut the last shingle so it fits tightly against the vertical shingle without overlapping it.

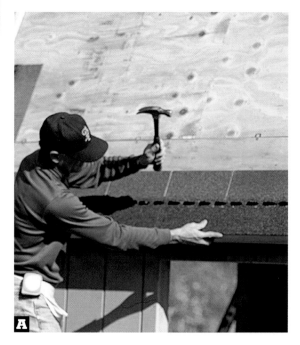

Shingle the roof

1. Start at one end of the roof and set the first architectural-style shingle directly on top of the starter course. Make sure its end and lower edge are flush with the starter shingle below, then fasten the shingle with four 1¼-in. roofing nails. Drive one nail 1 in. from each end of the shingle and space the other two nails 12 in. in between. Continue laying full-size shingles until you reach the end of the roof, being careful to overlap all the vertical seams in the starter course by at least 6 in. Then cut the last shingle to length, allowing for a ½-in. overhang at the end.

2. Start the second course by first cutting 6 in. off a shingle; that will automatically stagger the seams between the first and second course by 6 in. Place the first shingle in the second course flush with the end of the roof and align its lower edge to produce a 5-in. exposure to the weather, then secure it with four 1¼-in. roofing nails. The easiest way to maintain the correct exposure is to align the bottom edge of the shingles with the horizontal lines formed by the upper laminations on the shingles below.

Continue nailing down full-length shingles in this manner, as shown in **B**. As you make your way up the roof, add starter shingles, as needed, along the vertical ends of the roof.

3. To start the third course, cut 12 in. off a shingle; this 24-in.-long shingle will maintain the 6-in. offset between the seams. Repeat this pattern for each subsequent course. Start the fourth course with an 18-in. shingle, the fifth course with a 12-in. shingle, and the sixth course with a 6-in. shingle. Start the seventh course with a full-size shingle and begin the pattern all over again. When you get to the last

B

(uppermost) course, tuck the shingles underneath the upper-slope overhang, nailing them as close as possible to the fascia.

4. Use the same roofing technique to shingle the upper slopes. At the peak, trim the shingles flush with the edge of the plywood. Then install a flexible polyester-fiber ridge vent. Unroll the vent along the ridge and fasten it to the roof with 2-in.-long roofing nails. Don't pound the nails all the way in or you'll crush the vent. Cover the vent with hip-and-ridge shingles, making sure you maintain a 5-in. exposure to the weather. Fasten each shingle with two 2-in. roofing nails.

Windows and Exterior Trim

This barn has five 2-ft. by 3-ft. aluminum sliding windows that provide plenty of natural light and fresh air. There are two windows in each sidewall and one in the rear wall. The ready-to-install units have an integral mounting flange, insect

Many roofers snap chalklines while shingling to help keep shingle courses straight. These architectural-style shingles have a built-in alignment guide, and the bottom edge of each shingle is aligned flush with the horizontal edge of the upper lamination. Of course, this is all referenced from the first course; if that's off, then the whole roof will be off. I didn't snap chalklines, but I did measure up from the eave every few courses just to make sure I wasn't straying off course.

screens, and built-in weep holes that drain water trapped in the lower track. However, before you can set the windows in place, you must install the windowsills.

Install the windows

1. For each window, start by cutting a 44¼-in.-long windowsill from a 5/4 by 6-in. cedar board. Then cut a 4⅛-in.-wide by 4-in.-deep notch in each end.

2. Set the sill in the window opening and, if necessary, shim it level. Attach it to the 2×4 rough sill with six 2½-in.-long siding nails, as shown in **A**.

3. Set each aluminum window in its opening and press it tightly against the siding, as shown in **B**. As you hold it in place, have a helper inside the barn slide the window tightly against the wall stud on one side of the opening and secure it by driving two 2-in. screws through the mounting holes in the window's side jambs and into the wall stud. Slip a thin wood shim between the opposite side of the window and the wall stud and secure it with 2-in. screws.

Trim out the windows

1. Cut two 1×4 cedar side casings to 24¾ in. long. Set each one on top of the sill, making sure it overlaps the aluminum window frame by ½ in.

2. Attach the casings to the sidewall with 1½-in. siding nails, as shown in **C**. Cut a 1×4 cedar head casing to 43¼ in. long and set it on top of the side casings. Make sure it overhangs each casing an equal amount, then nail it in place.

3. Next, make the apron, which is the horizontal trim piece running underneath the sill. Cut a 1×4 cedar board to 43¼ in. long and nail it to the sidewall below the window-sill, as shown in **D**. Make sure the ends of the apron align with the outside edges of the side casings.

Add the stone floor

Before installing the doors, order a truckload of processed stone from a masonry supplier. You'll need between 4 cu. yd. and 6 cu. yd. Have the load dumped right at the doorway opening, then use a wheelbarrow to transport the stone into the barn. Put down a 4-in. to 6-in. layer of stone and rake it smooth.

Door Installation

The doors (which are shown in the sidebar on p. 210) are hung with traditional sliding-door hardware. Two roller assemblies are mounted on the top of each door. Each assembly has two round rollers. A round tubular steel track is bolted over the doorway opening, then the rollers are slipped into the track. This type of

hardware is called a cannonball hanger. Trolley hangers, which have wheels similar to those on roller skates, are also popular.

Attach the sliding-door hardware

1. Start by cutting two ⅞-in.-wide by 1½-in.-deep slots in the top of each door. Position the notches 5½ in. from the edges of the doors. Cut the notches using a handsaw, a hammer, and a sharp chisel.

2. Hold the rear mounting plate against the back of the door and bore two ⅜-in.-dia. holes through the doorframe. Pass the two carriage bolts through the holes. Insert the square nut from one of the roller assemblies into the notch cut in the door.

3. Place the front mounting plate on the door, trapping the square nut in the notch. Thread hex nuts onto the bolts and tighten them with a socket wrench, as shown in **A**.

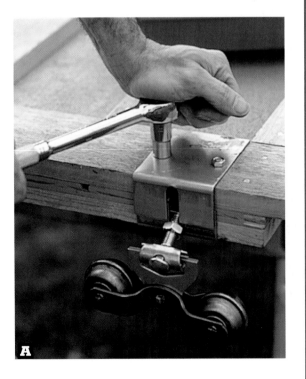

PRO TIP

If you're going to paint or stain the window trim, save yourself some time—and aggravation—by applying the finish before you install the trim.

WHAT CAN GO WRONG

Whenever you build doors out of plywood, there's always a chance that they may warp. That's because the cross-grain plies absorb moisture and expand and contract at different rates. To deter warping, apply at least two coats of paint or stain to all surfaces of the plywood panel, especially the porous edges. Reinforce the front and rear of the door with a wood frame made from redwood or cedar. Also, keep the space beneath the door free of debris that can trap moisture.

Building Sliding Doors

THE SLIDING DOORS ON THIS SHED resemble board-and-batten doors, but they are actually made from grooved plywood siding. Each 34-in. by 77-in. plywood panel is sandwiched between a front and a rear perimeter frame, then three 1×2 vertical battens are applied to the front surface. Cross battens, cut from 1×6 cedar, are screwed to the backs of the doors to add strength and deter warping. Here are the basic steps for building each door:

1. Cut the plywood door panel to size, then cut the four parts of the perimeter face frame from roughsawn cedar 1×4s. Cut two 35½-in.-long horizontal rails and two 70-in.-long vertical stiles.

2. Fasten the face frame to the plywood panel with construction adhesive and 1¼-in. (3d) galvanized nails, as shown in **A**. Note that the rails overhang the edge of the panel by 1½ in. to create an overlapping wind-stop joint where the two doors meet. Create a similar overlap on the edge of the mating door.

3. Use adhesive and nails to attach the three vertical 1×2 cedar battens to the door face, as shown in **B**. Center each batten directly over one of the grooves milled in the plywood.

4. Flip over the door. Rip five cedar 1×6s to 4½ in. wide, as shown in **C**. From these boards, cut two 34-in.-long rails, two 68-in.-long stiles, and one 25-in.-long center rail. Fasten the parts to the perimeter of the door panel with panel adhesive and 1¾-in. decking screws spaced no more than 10 in. apart.

5. Cut diagonal cross battens to fit between the horizontal rails screwed to the rear of the door. Attach the battens with adhesive and 1¾-in. decking screws, as shown in **D**.

A

B

C

D

grade beam. Make sure the screw passes through the plywood siding and into the 2×4 wall frame.

6. If there's no framing member directly behind the door track, add blocking. Use a 4-ft. level to make sure the track is perfectly level, then drive a lag screw through the mounting clip at the opposite end of the track. Check the track for level one more time, then screw the remaining four mounting clips to the wall.

Hang the doors

These doors are very easy to hang, but it takes two people to get them up onto the track. Start by setting up a ladder near the end of the tubular steel track. With the help of a friend, bring over one of the doors and lean it against the front wall of the barn. Climb up the ladder.

1. As your helper raises the door from below, guide the roller assemblies into the track, as shown in **D**. Slide the door all the way to the other end of the track and repeat the process to hang the second door.

4. Next, use a hacksaw to cut 12 in. off the 12-ft.-long tubular steel track. Note that the track has a series of keyhole-shaped slots cut in it. Insert a metal mounting clip in each of the six slots, as shown in **B**, using a hammer to gently tap the clip all the way in.

5. Raise the track over the doorway, as shown in **C**, and fasten one end by driving a ⅜-in.-dia. by 2-in.-long lag screw through the hole in the first mounting clip. The center of the rail should be 76½ in. above the top of the 6×6

2. Check to make sure the doors slide smoothly and don't rub against the barn. Also, bring the doors together to see whether they meet squarely. You may have to raise or lower one of the elevation-adjustment nuts on the roller assemblies to get the doors to hang straight and meet flush. If a door rubs against the siding, loosen the lateral-adjustment nuts on the roller assemblies. Use a hammer to tap the interior mounting plate until the door moves away from the barn. Tighten the nuts and test the door.

3. Once you're satisfied with the way the doors open and close, install the two lower guide-ball brackets. These pieces prevent the bottoms of the doors from swinging away from the barn. One bracket is needed for each door.

4. Open the doors all the way until they're flush with the edge of the doorway opening. Mount a guide-ball bracket 6 in. from the outer edge of each door, fastening the brackets to the 6×6 grade beam with ⅜-in.-dia. by 2-in. lag screws, as shown in **E**. Loosen the nut on the underside of the bracket and move the guide ball to within ⅛ in. of the door, then tighten the nut.

5. Finally, install 1×4 roughsawn cedar corner boards on the four corners of the barn. Miter-cut the top ends of the boards to match the slope of the roof and fasten them with 2½-in.-long siding nails spaced 12 in. apart, as shown in **F**.

Note that the corner boards can't be installed earlier because they will interfere with sliding the roller assemblies into the tubular steel track.

E

F

Bonus Storage

ONE BENEFIT OF BUILDING THIS STORAGE barn is that its double-sloping gambrel roof offers plenty of overhead storage space. Above the ceiling joists is an "attic" area that's about 6 ft. high by 6½ ft. wide by 18 ft. long. It's an ideal out-of-the-way place for storing seldom-used items, such as old tools, camping gear, holiday decorations, suitcases, and boxes of seasonal clothing.

After installing a plywood floor, the challenge is finding the best way to access the space. A fold-down attic staircase is a safer and more secure alternative than a simple ladder. The cleverly designed folding staircase is mounted on a pull-down door panel equipped with spring-loaded brackets. The hinged stairs fold flat against the panel and the brackets extend when the unit is lowered. To access the storage space, you simply pull down the door panel and unfold the stairs.

A fold-down attic staircase provides an easy and safe way to gain access to the overhead storage space above the ceiling joists.

RESOURCES

BUILDING PLANS FOR SHEDS SHOWN IN THIS BOOK

To order a set of building plans for the Lean-to Shed Locker, send a check for $19.95 to:

RWS Company
105 South St.
Roxbury, CT 06783
860-350-1089

To order a set of building plans for the Saltbox Potting Shed, Garden Shed, Colonial-Style Shed, or Gambrel Storage Barn, send a check for $29.95 per plan to:

Better Barns
126 Main St. South
Bethlehem, CT 06751
888-266-1960
www.betterbarns.com

OTHER SOURCES FOR MAIL-ORDER SHED PLANS

Handy Home Products
6400 E. 11 Mile Rd.
Warren, MI 48091
800-221-1849
www.handyhome.com

Barn Plans
213 Second Texas Rd.
St. George, SC 29477
843-563-7267
www.barnplan.com

Georgia-Pacific
133 Peachtree St., N.E.
Atlanta, GA 30303
404-652-4000
www.gp.com

For plans of the post-and-beam shed shown at right on p. 10, send a check for $75 to:

Mercurial Editorial
P.O. Box 94
Maryknoll, NY 10545
mmmedit@aol.com

PREBUILT SHED KITS

Lancaster County Barns
218-B West Main St.
Leola, PA 17540
717-556-0394
www.lancasterbarns.com

Outback Wood Products
9630 Black Mountain Rd.
San Diego, CA 92126
800-561-1193
858-695-0488
www.outbackwoodproducts.com

The Cedar Store
5410 Route 8
Gibsonia, PA 15044
888-293-2339
www.cedarstore.com

Backyard City
5910 Mt. Moriah
No. 113-240
Memphis, TN 38115
877-233-9829
www.backyardcity.com

Summerwood Inc.
733 Progress Ave.
Toronto, Ontario M1H 2W7
800-663-5042
www.summerwood.com

Walpole Woodworkers
767 East St.
Walpole, MA 02308
800-343-6948
www.walpolewoodworkers.com

Quality Outdoor Structures
866-324-2625
www.qualityoutdoorstructures.com

SHED HARDWARE

Sliding Door Hardware and Swinging Door Hinges and Latches

Better Barns
126 Main St. South
Bethlehem, CT 06751
888-266-1960
www.betterbarns.com

Sliding Barn Door Hardware

CannonBall: HNP
P.O. Box 835
Beloit, WI 53512
800-766-2825
www.cnbhnp.com

Gambrel-Roof Framing Hardware

Lee Valley Tools
P.O. Box 1780
Ogdensburg, NY 13669
800-871-8158
www.leevalley.com
Fast-Framer Kit, Item AC503

Shingle Shield Protector Strips

Chicago Metallic
4849 Austin Ave.
Chicago, IL 60638
800-323-7164
www.chicago-metallic.com

Metal Framing Connectors

Simpson Strong-Tie
P.O. Box 1568
San Leandro, CA 94577
800-999-5099
www.strongtie.com

PRESSURE-TREATED LUMBER

Arch Wood Protection
1955 Lake Park Dr., Suite 250
Smyrna, GA 30080
770-801-6600
www.wolmanizedwood.com

Chemical Specialties
200 E. Woodlawn Rd., Suite 250
Charlotte, NC 28217
800-421-8661
www.treatedwood.com

Georgia-Pacific
133 Peachtree St., N.E.
Atlanta, GA 30303
404-652-4000
www.gp.com

Osmose Wood Preserving
980 Ellicott St.
Buffalo, NY 14209
716-882-5905
www.osmose.com

Southern Pine Council
P.O. Box 641700
Kenner, LA 70064
504-443-4464
www.southernpine.com

CEDAR

Cedar Shake & Shingle Bureau
P.O. Box 1178
Sumas, WA 98295
604-820-7700
www.cedarbureau.org

Western Red Cedar Lumber Association
1200-555 Burrard St.
Vancouver, BC V7X 1S7
604-684-0266
www.wrcla.org

REDWOOD

California Redwood Association
405 Enfrente Dr., Suite 200
Novato, CA 94949
888-225-7339
www.calredwood.org

CONCRETE, MORTAR, AND MASONRY SUPPLIES

Bonsal
P.O. Box 241148
Charlotte, NC 28224
704-525-1621
www.bonsal.com

Sakrete
1402 N. River St.
Portland, OR 97227
800-245-3833
www.sakreteconcrete.com

Precast Dek-Block Pier Blocks

DekBrands
P.O. Box 14804
Minneapolis, MN 55414
952-746-7562
www.deckplans.com

MISCELLANEOUS SHED PRODUCTS

Ramp-Building Kit

Ramparts
Highland Group Industries
31200 Solon Rd., Suite 1
Solon, OH 44139
800-234-6992
www.ramparts.com

Cellular PVC Trim

Fypon
800-446-3040
www.fypon.com

Azek Trimboards
877-275-2935
www.azek.com

Fiber-Cement Siding

CertainTeed
P.O. Box 860
Valley Forge, PA 19482
610-341-7000
www.certainteed.com

James Hardie Building Products
26300 La Alameda, Suite 250
Mission Viejo, CA 92691
888-JHARDIE
www.jameshardie.com

Dura-Slate Roof Shingles

Royal Building Products
1 Royal Gate Blvd.
Woodbridge, Ontario L4L 8Z7
905-264-0701
www.royalbuildingproducts.com

Exterior Wood Finishes

Samuel Cabot Inc.
100 Hale St.
Newburyport, MA 01950
800-US-STAIN
www.cabotstain.com

Aluminum Sliding Windows

PGT Industries
1070 Technology Dr.
Nokomis, FL 34275
877-550-6006
www.pgtindustries.com

Solar-Powered Photovoltaic Cells

Real Goods
360 Interlocken Blvd., Suite 300
Broomfield, CO 80021
800-762-7325
www.realgoods.com

Solar-Powered Products

Solar Living Institute
P.O. Box 836
Hopland, CA 95449
707-744-2017
www.solarliving.org

INDEX